Exploration Activities

Mathematics for Elementary School Teachers
A Process Approach

Mark A. Freitag
Augusta State University

Prepared by

Mark A. Freitag
Augusta State University

Linda B. Crawford
Augusta State University

BROOKS/COLE
CENGAGE Learning

Australia • Brazil • Japan • Korea • Mexico • Singapore • Spain • United Kingdom • United States

BROOKS/COLE
CENGAGE Learning

© 2014 Brooks/Cole, Cengage Learning

ALL RIGHTS RESERVED. No part of this work covered by the copyright herein may be reproduced, transmitted, stored, or used in any form or by any means graphic, electronic, or mechanical, including but not limited to photocopying, recording, scanning, digitizing, taping, Web distribution, information networks, or information storage and retrieval systems, except as permitted under Section 107 or 108 of the 1976 United States Copyright Act, without the prior written permission of the publisher.

> For product information and technology assistance, contact us at
> **Cengage Learning Customer & Sales Support,
> 1-800-354-9706**
>
> For permission to use material from this text or product, submit all requests online at **www.cengage.com/permissions**
> Further permissions questions can be emailed to
> **permissionrequest@cengage.com**

ISBN-13: 978-1-133-96315-8
ISBN-10: 1-133-96315-3

Brooks/Cole
20 Davis Drive
Belmont, CA 94002-3098
USA

Cengage Learning is a leading provider of customized learning solutions with office locations around the globe, including Singapore, the United Kingdom, Australia, Mexico, Brazil, and Japan. Locate your local office at: **www.cengage.com/global**

Cengage Learning products are represented in Canada by Nelson Education, Ltd.

To learn more about Brooks/Cole, visit
www.cengage.com/brookscole

Purchase any of our products at your local college store or at our preferred online store
www.cengagebrain.com

Printed in the United States of America
1 2 3 4 5 6 7 16 15 14 13 12

Contents

Preface ix

Chapter 1 Mathematical Processes 1
   Exploration 1.1 Writing Your Own Mathography 3
   Exploration 1.2 Creating Designs from Directions 5
   Exploration 1.3 Patterns of Common Traits 7
   Exploration 1.4 Growing Patterns 9
   Exploration 1.5 Venn Diagrams and Valid Arguments 11
   Exploration 1.6 Lewis Carroll Logic Puzzles 13
   Exploration 1.7 The Sum of an Even and an Odd Number 15
   Exploration 1.8 The Sum of Three Consecutive Numbers 17
   Exploration 1.9 Cutting a Pie 18
   Exploration 1.10 Seating at a High School Graduation 19
   Exploration 1.11 How Much Do School Supplies Cost? 20
   Exploration 1.12 Humans and Horses 21

Chapter 2 Sets 23
   Exploration 2.1 An Introduction to Attribute Blocks 25
   Exploration 2.2 Difference Puzzles 27
   Exploration 2.3 Subsets, Equal Sets, and Equivalent Sets 31
   Exploration 2.4 Counting in the Elementary Classroom 33
   Exploration 2.5 Attribute Blocks and Set Operations 35
   Exploration 2.6 Sorting by Two and Three Attributes 37

Chapter 3 Numbers and Numeration 41
   Exploration 3.1 Exploring Mayan Numeration 43
   Exploration 3.2 Efficiency in Different Numeration Systems 45
   Exploration 3.3 A Trip to Hexaminia 47
   Exploration 3.4 Base-ten Blocks and the Decimal System 50
   Exploration 3.5 Conceptualizing Large Numbers 53
   Exploration 3.6 Understanding Base-b Numerals 55

Chapter 4 Whole-Number Computation 57
   Exploration 4.1 Properties of Whole-Number Addition 59
   Exploration 4.2 Addition and Subtraction Story Problems 62
   Exploration 4.3 Addition with Base-Ten Blocks 64
   Exploration 4.4 Subtraction with Base-Ten Blocks 66
   Exploration 4.5 Mental Addition and Subtraction 68
   Exploration 4.6 Children's Strategies for Addition and Subtraction 69

Exploration 4.7 Children's Errors with Addition and Subtraction … 71
Exploration 4.8 Properties of Whole-Number Multiplication … 73
Exploration 4.9 Multiplication and Division Story Problems … 76
Exploration 4.10 Interpreting Remainders in Division … 78
Exploration 4.11 Multiplication with Base-Ten Blocks … 80
Exploration 4.12 Division with Base-Ten Blocks … 82
Exploration 4.13 Mental Multiplication and Division … 84
Exploration 4.14 Children's Strategies for Multiplication and Division … 85
Exploration 4.15 Children's Errors with Multiplication and Division … 88
Exploration 4.16 Hexaminian Addition and Subtraction … 90
Exploration 4.17 Hexaminian Multiplication and Division … 93

Chapter 5 Basic Number Theory … 97
Exploration 5.1 Making Sense of Divisibility by 2, 5, and 10 … 99
Exploration 5.2 Making Sense of Divisibility by 4 and 8 … 101
Exploration 5.3 Making Sense of Divisibility by 3 and 9 … 103
Exploration 5.4 The Factor Game … 105
Exploration 5.5 Finding Primes to 200 … 109
Exploration 5.6 Understanding the Prime Factorization Method for Finding the GCF and LCM … 111
Exploration 5.7 Understanding the Euclidean Algorithm … 113
Exploration 5.8 Problems Involving Cycles … 115

Chapter 6 The Integers … 117
Exploration 6.1 Integers and the Chip Model … 119
Exploration 6.2 Integers and the Number Line Model … 122
Exploration 6.3 Integer Addition and Subtraction with the Chip Model … 125
Exploration 6.4 Integer Addition and Subtraction on a Number Line … 127
Exploration 6.5 Modeling Integer Multiplication … 129
Exploration 6.6 Exploring Integer Division … 131

Chapter 7 Fractions and the Rational Numbers … 133
Exploration 7.1 Exploring Fractions with the Area Model … 135
Exploration 7.2 Exploring Fractions with the Length Model … 138
Exploration 7.3 Exploring Fractions with the Set Model … 141
Exploration 7.4 Student Errors in Understanding the Fractions … 143
Exploration 7.5 Why is a Common Denominator Necessary? … 145
Exploration 7.6 Adding and Subtracting Fractions … 147
Exploration 7.7 Fraction Sense with Addition and Subtraction … 149
Exploration 7.8 Student Errors in Adding and Subtracting Fractions … 150
Exploration 7.9 The Meaning of Fraction Multiplication … 152

Contents v

Exploration 7.10 The Meaning of Fraction Division — 154
Exploration 7.11 Multiplying and Dividing Fractions — 157
Exploration 7.12 Fraction Sense with Multiplication and Division — 160
Exploration 7.13 Computing with a Fraction Calculator — 161
Exploration 7.14 Student Errors in Multiplying and Dividing Fractions — 163

Chapter 8 Decimals, Real Numbers, and Proportional Reasoning — 165

Exploration 8.1 Base-Ten Blocks and Decimal Numbers — 167
Exploration 8.2 Modeling Decimal Numbers — 169
Exploration 8.3 Understanding Repeating Decimals and Irrational Numbers — 172
Exploration 8.4 Where is One for You? — 174
Exploration 8.5 Multiplying Decimals — 181
Exploration 8.6 Connecting Decimal Operations to Fraction Operations — 183
Exploration 8.7 Student Errors in Using Scientific Notation — 185
Exploration 8.8 Understanding Proportional Reasoning Problems — 187
Exploration 8.9 Solving Proportion Problems with the Bar Model — 189
Exploration 8.10 The Golden Ratio and the Human Body — 191
Exploration 8.11 Student Errors in Proportional Reasoning — 193
Exploration 8.12 Solving Percent Problems with the Bar Model — 194
Exploration 8.13 Grading with Points and Percents — 196

Chapter 9 Algebraic Thinking — 199

Exploration 9.1 Understanding Arithmetic Sequences — 201
Exploration 9.2 Understanding Geometric Sequences — 204
Exploration 9.3 The Fibonacci Sequence and the Golden Ratio — 207
Exploration 9.4 Understanding the Meaning of a Function — 209
Exploration 9.5 Functional Thinking and Patterns of Change — 211
Exploration 9.6 Functions in the Elementary Classroom — 213
Exploration 9.7 Graphing Functions on a Calculator — 217
Exploration 9.8 Exploring the Slope and Y-Intercept of a Line — 219
Exploration 9.9 Arithmetic Sequences and Linear Functions — 221
Exploration 9.10 Solving Linear Equations with Algebra Tiles — 223
Exploration 9.11 Student Errors in Solving Linear Equations and Inequalities — 226
Exploration 9.12 Mathematical Modeling and Functions — 227
Exploration 9.13 Mathematical Modeling and Equations — 229

Chapter 10 Geometrical Shapes — 231

Exploration 10.1 Measuring Angles with Manipulative Pieces — 233
Exploration 10.2 Segments and Angles on a Geoboard — 235
Exploration 10.3 Turtle Geometry — 238

Exploration 10.4 Sorting Planar Shapes — 240
Exploration 10.5 The Sides and Angles of a Triangle — 245
Exploration 10.6 Angle Measures in a Triangle — 248
Exploration 10.7 Making Sense of the Pythagorean Theorem — 250
Exploration 10.8 Properties of Quadrilaterals — 253
Exploration 10.9 Paper Folding and Polygons — 256
Exploration 10.10 Spatial Reasoning with Manipulatives — 258
Exploration 10.11 Angle Properties of Polygons — 260
Exploration 10.12 Spatial Reasoning in Three Dimensions — 265
Exploration 10.13 Three-Dimensional Shapes and Their Nets — 268
Exploration 10.14 Regular and Semi-Regular Polyhedra — 270
Exploration 10.15 Euler's Formula — 272

Chapter 11 Congruence, Similarity, and Constructions — 275

Exploration 11.1 Congruent Triangles and Conditions for Congruency — 277
Exploration 11.2 Congruent Shapes — 279
Exploration 11.3 Measuring with Similar Triangles — 282
Exploration 11.4 Similar Shapes — 284
Exploration 11.5 Rep-tiles and Fractals — 287
Exploration 11.6 Constructing Shapes With a Plastic Reflector — 289
Exploration 11.7 Basic Constructions with Geometry Software — 291
Exploration 11.8 Constructing Parallel and Perpendicular Lines — 293
Exploration 11.9 Centers of Triangles — 295
Exploration 11.10 Paper Folding and Constructions — 297
Exploration 11.11 Constructing Triangles and Quadrilaterals — 299
Exploration 11.12 Constructing Regular Polygons — 301

Chapter 12 Coordinate and Transformation Geometry — 303

Exploration 12.1 Taxicab Geometry — 305
Exploration 12.2 Coordinate Geometry on a Geoboard — 307
Exploration 12.3 Coordinate Geometry and Polygons — 309
Exploration 12.4 Making a Coordinate Map — 311
Exploration 12.5 Making Transformations — 313
Exploration 12.6 Exploring Transformations with Geometry Software — 315
Exploration 12.7 Spatial Reasoning and Transformations — 317
Exploration 12.8 Size Transformations — 319
Exploration 12.9 Symmetry — 321
Exploration 12.10 Defining Shapes with Symmetry — 323
Exploration 12.11 Border Patterns — 324
Exploration 12.12 Wallpaper Patterns — 326
Exploration 12.13 Making Tessellations — 328

Chapter 13 Measurement — **331**
    Exploration 13.1 Nonstandard Units and Measures of Length — 333
    Exploration 13.2 Understanding Metric Lengths — 335
    Exploration 13.3 Conversions Between English and Metric Lengths — 336
    Exploration 13.4 Estimating Distances — 338
    Exploration 13.5 Measuring Perimeters with the String Technique — 340
    Exploration 13.6 Perimeter and Area—Are They Related? — 342
    Exploration 13.7 Misconceptions about Perimeter and Area — 344
    Exploration 13.8 Pick's Theorem — 346
    Exploration 13.9 Area Formulas of Polygons — 348
    Exploration 13.10 Finding the Area of a Circle — 351
    Exploration 13.11 Understanding and Measuring Surface Area — 352
    Exploration 13.12 Understanding and Measuring Volume — 354
    Exploration 13.13 Estimating the Volume of Ice Cream — 357
    Exploration 13.14 The Volume of a Sphere — 359
    Exploration 13.15 Measuring Weight and Speed — 361
    Exploration 13.16 Measuring Temperature with Linear Functions — 363

Chapter 14 Statistical Thinking — **365**
    Exploration 14.1 Selecting Samples of Education Majors — 367
    Exploration 14.2 Writing and Conducting a Survey — 371
    Exploration 14.3 Collecting Data Through Observations — 372
    Exploration 14.4 The U.S. Census Bureau — 373
    Exploration 14.5 Data Collection in the Classroom — 374
    Exploration 14.6 Representing a Data Set with Multiple Graphs — 375
    Exploration 14.7 Making Statistical Graphs with a Spreadsheet — 377
    Exploration 14.8 Correlation and the Line of Best Fit — 379
    Exploration 14.9 Understanding Graphs with Percents — 381
    Exploration 14.10 Understanding Measures of Center — 383
    Exploration 14.11 Understanding Box Plots — 385
    Exploration 14.12 Understanding the Standard Deviation — 388
    Exploration 14.13 Understanding the Normal Distribution — 391
    Exploration 14.14 Percentiles — 393
    Exploration 14.15 Misleading Statistical Graphs — 395

Chapter 15 Probability — **397**
    Exploration 15.1 Probability as a Measure of Chance — 399
    Exploration 15.2 Testing the Law of Large Numbers — 401
    Exploration 15.3 What's in the Bag? — 403
    Exploration 15.4 Simulations — 404

Exploration 15.5 Experimental and Theoretical Probabilities — 407
Exploration 15.6 Properties of Probability — 409
Exploration 15.7 Multi-Stage Probability — 412
Exploration 15.8 Misconceptions with Probability — 414
Exploration 15.9 Geometric Probability — 416
Exploration 15.10 Understanding Odds — 418
Exploration 15.11 Fair Games — 420
Exploration 15.12 Pascal's Triangle and Combinations — 422

Appendix A: Black Line Masters — A-1

Decimal Place Value Mat — A-2
Equation Mat — A-3
One-Inch Grid Paper — A-4
Half-Inch Grid Paper — A-5
Quarter-Inch Grid Paper — A-6
Decimal Grid Paper — A-7
Rectangular Dot Paper — A-8
Isometric Dot Paper — A-9
5 × 5 Geoboard Paper — A-10
6 × 6 Geoboard Paper — A-11
11 × 11 Geoboard Paper — A-12
Table of Random Digits — A-13

Appendix B: Manipulatives — B-1

Algebra Tiles — B-2
Attribute Blocks — B-3
Base-Ten Blocks — B-7
Colored Rods — B-10
Counters — B-11
Fraction Bars — B-12
Fraction Disks — B-13
One-Inch Square Tiles — B-14
Pattern Blocks — B-15
Pentominoes — B-18
Tangrams — B-19
Two-Colored Chips — B-20

Preface

Welcome to the explorations manual for the text *Mathematics for Elementary School Teachers: A Process Approach*. This book serves as a companion to the text by providing a hands-on process approach for helping preservice teachers engage and understand the mathematics they are likely teach. It provides opportunities to explore mathematical ideas and to develop a conceptual understanding of mathematics by using a variety of investigations, materials, and technology. It also gives an excellent way to demonstrate an alternative approach to the traditional teaching and learning of mathematics.

How to Use This Manual

The explorations in this manual were developed under the philosophy that students learn better when they engage content through hands-on activities centered around worthwhile mathematical tasks. Through these explorations students will think through important mathematical concepts, procedures, and structures, reinforce concepts they may have been previously taught, or work with applications of the content relevant to the elementary classroom or other real-world situations.

By completing these activities, students will not only gain a deeper understanding of the mathematics they are likely to teach, but they will also encounter different approaches to working with the material that will help them be more effective teachers. Like the main text, this manual uses a process approach as its central pedagogical theme; that is, the explorations are designed to engage students in communicating mathematically, solving problems, reasoning mathematically, using representations, and making connections. Many of the tasks and questions are specifically designed to achieve the objectives associated with these processes as stated in the *NCTM Principles and Standards for School Mathematics*. These objectives are summarized in Table 1. By using this process approach, preservice teachers will not only come to understand the different processes, but also appreciate them as integral to learning and doing mathematics. This broader view will enable preservice teachers to give their students a well-rounded and holistic view of mathematics once they enter the classroom.

This explorations manual is not intended to be a textbook for a mathematics content course or to provide all of the mathematical content necessary for such a course. Rather, the explorations in this manual are designed to supplement classroom instruction. They can be used to reinforce concepts that have been covered in class, or they can be used to replace traditional lectures and still cover content relevant to the course. The explorations are appropriate for use with preservice teachers taking mathematics content courses or methods courses, or they can be used with inservice teachers taking professional development courses. Many of the explorations can also be adapted for use with students in the elementary and middle grades.

NCTM Process Standard	Instructional programs from prekindergarten through grade 12 should enable all students to:
Communication Standard	• Organize and consolidate their mathematical thinking through communication. • Communicate their mathematical thinking coherently and clearly to peers, teachers, and others. • Analyze and evaluate the mathematical thinking and strategies of others. • Use the language of mathematics to express mathematical ideas precisely.
Representation Standard	• Create and use representations to organize, record, and communicate mathematical ideas. • Select, apply, and translate among mathematical representations to solve problems. • Use representations to model and interpret physical, social, and mathematical phenomena.
Reasoning and Proof Standard	• Recognize reasoning and proof as fundamental aspects of mathematics. • Make and investigate mathematical conjectures. • Develop and evaluate mathematical arguments and proofs. • Select and use various types of reasoning and methods of proof.
Problem Solving Standard	• Build new mathematical knowledge through problem solving. • Solve problems that arise in mathematics and in other contexts. • Apply and adapt a variety of appropriate strategies to solve problems. • Monitor and reflect on the process of mathematical problem solving.
Connections Standard	• Recognize and use connections among mathematical ideas. • Understand how mathematical ideas interconnect and build on one another to produce a coherent whole. • Recognize and apply mathematics in contexts outside of mathematics.

Source: NCTM STANDARDS Copyright © 2011 by NATIONAL COUNCIL OF TEACHERS OF MATHEMATICS. Reproduced with permission of NATIONAL COUNCIL OF TEACHERS OF MATHEMATICS.

The manual contains a variety of explorations for each section. Some explorations deal directly with the content of the chapter, often making use of relevant manipulatives to explore the content. Other explorations extend the content of the section either mathematically or by building a connection to the K–8 classroom. In general, the order of the explorations matches the order of the content in the main text. However, the explorations are independent of one another, so they can be rearranged if necessary to better fit a particular instructor's classroom needs. Instructors should also feel free to select explorations using their own discretion. Each exploration is referenced in the main text with an icon in the margin next to the relevant content.

Each exploration begins with the purpose and a materials list. The **Purpose** states the learning objectives for the exploration by outlining the mathematical concepts, procedures, or connections that are developed in the exploration. The **Materials** list gives the handouts, manipulatives, or technology needed to complete the exploration. Most of the handouts required by the labs are provided in the manual. Some are unique to specific explorations and follow directly after the exploration. Other handouts are required for several explorations. These handouts can be found in the rear of the book in Appendix A. They can be torn out and permission is given for photocopying these handouts as necessary. The handouts that appear in Appendix A include

- A decimal place value mat
- 1-in. grid paper
- Quarter-inch grid paper
- Isometric dot paper
- 5 × 5 geoboard dot paper
- 11 × 11 geoboard dot paper
- An Equation mat
- Half-inch grid paper
- Decimal grid paper
- Rectangular dot paper
- 6 × 6 geoboard dot paper
- A table of random digits

A variety of manipulatives common to the elementary and middle grades are also used for the explorations. To save on expenses, many of them are provided as paper cut-outs in Appendix B. Specifically, Appendix B contains

- Algebra tiles
- Base-ten blocks
- Counters
- Fraction disks
- Pattern blocks
- Tangrams
- Attribute blocks
- Colored rods
- Fraction bars
- One-Inch Square Tiles
- Pentominoes
- Two-colored chips

Some explorations also incorporate relevant technology. Calculators and computer software that are used throughout this manual include fraction, scientific, and graphing calculators, a word processor, a spreadsheet, dynamic geometry software such as Geometer's Sketchpad®, and Internet access.

Other materials used throughout the explorations include:

- Boxes
- Cardstock
- Circular lids or objects
- Coffee stirrers
- Colored paper
- Colored pencils
- Compass
- Connecting Cubes
- Deck of cards
- Dice of various kinds
- Dried beans
- Fettuccine
- Fillable geometric solids
- Funnel
- Geometric shapes
- Half-gallon of Ice cream
- Modeling clay
- News papers or magazines
- One-inch colored cubes
- Overhead transparencies
- Pan or tray
- Paper bags
- Paper clips
- Patty paper
- Pennies or other coins
- Plastic figures
- Plastic reflectors
- Protractors
- Popsicle sticks
- Rice
- Rubber bands
- Rulers
- Scales
- Scissors
- Spinners
- State curriculum standards
- Stop watches
- String, various lengths
- Sugar ice cream cones
- Tape
- Tape measures
- Thermometers, Celsius and Fahrenheit
- Yard sticks and meter sticks

The body of each exploration is divided into two to five parts that follow in a sequential order to guide students through the objectives of the exploration. Each part provides directions for completing the activity, and in many cases, a short explanation of the relevant mathematical concepts or procedures is given. Students are then asked to answer a number of questions that guide them in exploring the mathematics and that encourage them to communicate mathematically, use representations, reason, solve problems, and build connections. Many of the questions are open-ended; that is, they can have more than one correct answer.

Although many of the explorations can be done individually, it is strongly recommended that students work cooperatively in small groups of two or three to complete the explorations. Since the use of small group learning in K-12 education is highly encouraged, preservice teachers need to experience the benefits of small group learning for themselves. Those benefits apply to all students, no matter their age. Research offers strong evidence that students must *construct* their own mathematical understanding in order to learn mathematics well. To construct their own mathematical understanding, students need opportunities to examine, represent, transform, solve, apply, prove, and communicate within the mathematics they are learning. These opportunities occur most readily when students work in small groups where they can exchange ideas, clarify their own thinking, support or challenge the thinking of others, find solutions, and generalize results. In many senses, working in small groups allows students to take charge of their own learning.

Putting students in groups is no guarantee that the groups will function well. To improve how the groups work, you may want to ask students to reflect on the communication and interaction within groups. Some questions to facilitate this reflection may include:

- How did I contribute to the successful operation of my group?
- What can I do to help the group function better?
- Did the group members ask questions when they did not understand?
- Did everyone have a chance to contribute ideas?
- Did the group members listen to one another?
- Did any one person take over the group?

Each exploration concludes with a **Summary and Connections** feature. These questions ask students to summarize or extend what they have learned. They may also have students build connections between the content of the exploration and their future classrooms.

Each exploration should take 30–45 minutes to complete. After an activity is completed, it is recommended that the instructor facilitate a discussion that summarizes the activity and reviews the mathematical concepts developed in the exploration. These discussions can also provide an opportunity to assess students' mastery of the content.

Content in the Explorations

The content of this explorations manual was developed to correspond to the content in the text *Mathematics for Elementary School Teachers: A Process Approach*. The text uses the five Content Standards from the *NCTM Principles and Standards for School Mathematics* as its general framework. Specifically:

- The Number and Operations Standard is addressed in Chapters 2–8.
- The Algebra Standard is addressed in Chapter 9.
- The Geometry Standard is addressed in Chapters 10–12.
- The Measurement Standard is addressed in Chapter 13.
- The Data Analysis and Probability Standard is addressed in Chapters 14 and 15.

The chapters in this exploration manual follow the same general framework. Not every concept or procedure in the main text is addressed in this manual. Rather, most of the explorations focus on the key concepts from each chapter. The following list provides a summary of how the key concepts from each chapter are addressed in this manual.

Chapter 1 Mathematical Processes

Chapter 1 introduces the five processes that play an important role in learning mathematics: communication, representation, reasoning and proof, problem solving, and connections. Four of the processes involve important mathematical ideas in their own right, so they are discussed in further detail throughout the chapter. The explorations for this chapter provide an additional opportunity to engage students in using the processes. The first two explorations focus on mathematical communication by having students write their own mathography and create designs from directions. The next explorations concentrate on reasoning; some use inductive reasoning to explore patterns of common traits and growing patterns while others use deductive reasoning to verify conjectures. The final explorations of the chapter use Polya's problem-solving method to solve a variety of problems.

Chapter 2 Sets

Chapter 2 begins the content on number and numeration by introducing sets. Although not formally taught in the K–8 curriculum, sets and set operations form the foundation for much of elementary mathematics. Many of the explorations in this chapter use attribute blocks to explore these ideas. Specifically, attribute blocks are used to investigate set relationships and operations and to solve sorting problems. One other exploration also considers the different types of counting often used in the elementary grades.

Chapter 3 Numbers and Numeration

The explorations in Chapter 3 allow students to explore many of the important ideas associated with numeration, such as place value and bases. Some explorations investigate these ideas by working with them in different numeration systems. Other explorations are designed to develop number sense. Base-ten blocks and other common manipulatives are used throughout.

Chapter 4 Whole-Number Computation

Chapter 4 explorations help students develop many of the concepts surrounding the whole-number operations of addition, subtraction, multiplication, and division. In some explorations students will consider the meaning of the operations, their properties, and the contexts in which they are used. In others they will investigate computations with large numbers and the strategies and mistakes children are likely to make while performing them. Base-ten blocks and other common manipulatives are used throughout.

Chapter 5 Basic Number Theory

Chapter 5 introduces notions from basic number theory. The explorations in this chapter take an in-depth look at divisibility tests, investigate factors and prime numbers, and use different methods to compute the greatest common factor and the least common multiple.

Chapter 6 The Integers

Chapter 6 expands the set of whole numbers to the set of integers. The first four explorations use the chip and number line models to represent the integers and develop integer addition and subtraction. The last two explorations examine the meaning of integer multiplication and division.

Chapter 7 Fractions and the Rational Numbers

Chapter 7 expands the set of integers to the set of rational numbers expressed in fractional form. The first several explorations use different manipulatives like fractions bars, fraction disks, and pattern blocks to develop an intuitive understanding of the rational numbers and their properties. Subsequent explorations then consider rational number operations and the misconceptions children have in performing them.

Chapter 8 Decimals, Real Numbers, and Proportional Reasoning

Chapter 8 concludes the content on number sets by developing the decimals and the set of the real numbers. Specifically, the first three explorations consider the fundamental ideas of the decimals and different ways to represent them. The next four then examine decimal operations and their connection to whole-number and rational number operations. The last several explorations address proportional reasoning.

Chapter 9 Algebraic Thinking

The three key components of algebraic thinking relevant to the K–8 curriculum are discussed in Chapter 9. These components are functional thinking, generalized arithmetic, and mathematical modeling. The first three explorations use numerical sequences to develop functional thinking. The five explorations that follow continue to work on functional thinking by examining ways of representing functions and working with linear functions. For generalized arithmetic, the chapter has two explorations on solving equations and the common mistakes students make. The chapter ends with two explorations that use mathematical modeling to solve problems.

Chapter 10 Geometrical Shapes

Chapter 10 begins a investigation of geometrical shapes and their properties. The opening explorations use manipulatives and turtle geometry to investigate lines, planes, and angles. The next set of explorations examines triangles, quadrilaterals, and other polygons for their characteristics and properties. The final four explorations analyze the properties of three-dimensional shapes.

Chapter 11 Congruence, Similarity, and Constructions

Chapter 11 continues the content on geometry by introducing congruence and similarity. The explorations in this chapter begin with considering congruent triangles and other shapes and then move on to similar shapes and self-similarity. The last seven explorations use a variety of tools and computer software to complete geometrical constructions.

Chapter 12 Coordinate and Transformation Geometry

Chapter 12 concludes the content on geometry by presenting two topics related to spatial reasoning: coordinate and transformation geometry. The first four explorations look at ways to represent shapes on the coordinate plane and then consider applications of coordinate geometry. The next set of explorations focuses on understanding transformations and their use in describing symmetry. The chapter concludes with three explorations that investigate ways to make and classify geometrical patterns.

Chapter 13 Measurement

Chapter 13 focuses on the basic concepts of measurement. The first four explorations examine measurement units and length. The remaining explorations investigate other measurable attributes including perimeter, area, surface area, volume, weight, time, and temperature.

Chapter 14 Statistical Thinking

Chapter 14 addresses statistical thinking with a consideration of the four basic steps included in most statistical studies. The first five explorations focus on tasks such as posing questions, selecting samples, and collecting data in different ways. In the next several explorations, statistical graphs are created and analyzed in a variety of ways. The chapter ends with explorations that investigate and analyze descriptive statistics.

Chapter 15 Probability

Ideas related to probability and its connection to data analysis compose the content of Chapter 15. The first set of explorations examines experimental probability and the Law of Large Numbers. The explorations then move on to investigating theoretical probability, its properties, and different ways to compute it. The last set of explorations considers other topics associated with probability such as odds, expected value, and counting techniques.

As your students work through the explorations contained in this manual, we hope they will not only gain a deeper understanding of the mathematics they are learning but will also enjoy the experiences and opportunities provided.

Chapter 1
Mathematical Processes

Elementary teachers play a particularly important role in mathematics education by helping students to develop a solid foundation for future studies in mathematics. They help students not only to understand mathematical content, but also to master the processes through which mathematics is done and learned.

Most modern mathematics curricula teach five processes: mathematical communication, representation, reasoning and proof, problem solving, and connections. The first four are independent of specific content; that is, they can be integrated and learned at any grade level and with any mathematical content. However, these processes involve important mathematical ideas in their own right, so they warrant individual attention and are the focus of Chapter 1. The fifth process, developing mathematical connections, is content specific, so it is difficult to talk about the process without using particular mathematical ideas. In upcoming chapters, many of the explorations will investigate how mathematics is connected within itself, to other disciplines, and to real-world situations.

The explorations in this chapter are designed to help you think about and work with the mathematical processes. Specifically,

- Explorations 1.1 - 1.2 concentrate on mathematical communication and representation.
- Explorations 1.3 - 1.8 focus on two important types of mathematical reasoning: inductive and deductive reasoning.
- Explorations 1.9 - 1.12 use Polya's problem-solving method to investigate and solve several problems.

Exploration 1.1 Writing Your Own Mathography

Purpose: Students explore their own views of mathematics and think about how these views might shape not only how they teach mathematics but also the views of their students towards mathematics.

Materials: Paper and pencil

Part A: Discovering Your Attitude Toward Mathematics

As a future elementary teacher, it is important to realize that the beliefs and attitudes that teachers communicate, even when unintentional, have a lasting impact on the beliefs and attitudes of students. Teachers who are enthusiastic about the content they teach can provide a supportive environment that can have a positive impact on their students' attitudes. Unfortunately, the reverse is also true. Teachers with a poor attitude about particular content can have a detrimental effect on their students' attitudes toward that same content.

The relationship between teacher and student attitudes applies to any subject but seems particularly pronounced in mathematics. Research suggests that many elementary teachers have a poor attitude towards mathematics, which can then be passed on to their students. In general, children are neither born with an inherent dislike or fear of mathematics, nor do they enter school having poor attitudes towards mathematics. Instead, they often learn such beliefs from their parents, their peers, or their teachers.

To make sure you portray a good attitude about mathematics in your classroom, it is important for you to recognize the attitudes and beliefs you currently hold about mathematics. The following questions are designed to help you reflect about your past experiences and how they have shaped your beliefs and attitudes towards mathematics. Once you have completed the questions, sit down with two or three of your peers and discuss what you have learned.

1. a. Describe an experience you had in a mathematics class that was good. What made the experience good? How did this affect your view of mathematics?
   b. Describe an experience you had in a mathematics class that was bad. What made the experience bad? How did it affect your view of mathematics?

2. Complete each sentence.
   a. The easiest thing I find about mathematics is …
   b. The hardest thing I find about mathematics is…

3. Draw a picture of what you think the typical mathematics teacher looks like. What does your picture reveal about your attitudes towards mathematics and those who teach it?

Part B: Mathematics Anxiety

In some cases, the attitudes people have about mathematics go beyond a simple dislike of the subject and develop into **mathematics anxiety**. This feeling of anxiety interferes with a person's ability to do and understand mathematics in ordinary life and in academic situations. It leads to feelings of helplessness and being out of control, which in turn can cause a person to avoid mathematical situations. Unfortunately, many students have their first encounter with these kinds of feelings in the late elementary grades.

Teacher behavior may be the single largest factor in causing mathematics anxiety. It is easy to see how overt behaviors such as showing frustration with student questions or pointing out a student's errors in front of the class can cause anxiety. More subtle behaviors, however, can be just as damaging. For instance, if a teacher gives preference to those that have more aptitude for math, then students who are not as quick are likely to feel inferior.

1. Make a list of the things a teacher might do that may cause students to dislike mathematics or cause them to have mathematics anxiety. Reflect on whether you have had experiences with teachers who have exhibited these types of bad behaviors. How did it make you feel about the mathematics you were learning?

2. What strategies might you use in the classroom to ensure that you are not instilling any negative attitudes or beliefs about mathematics in your students?

3. Another reason students become math anxious may have to do with some of the content itself. Think about the mathematical topics that are taught in the late elementary and early middle grades. What content might make children anxious while learning it? Why do you think this is the case?

Part C: Summary and Connections

1. Why is it reasonable to consider mathematics anxiety a part of mathematical communication as described in the opening section of the text?

2. Write a short paragraph that summarizes what this activity has helped you to discover about your own attitudes and beliefs about mathematics. How might this influence the way you teach mathematics in the future?

Exploration 1.2 Creating Designs from Directions

Purpose: Students follow directions to create a design with pattern blocks and then reflect on the communication involved in the activity.

Materials: Pattern blocks

Part A: Design 1

1. Use a set of pattern blocks to make a bird with
   - A hexagon for its body.
   - A square for its head.
   - Triangles for the beak and tail.
   - Triangles for the feet.

2. Compare your bird to one created by one of your peers.
   a. Do your birds match? If not, how do they differ?
   b. Is one more correct than the other based on the given directions? How do you know?

3. Can you use the same set of directions to make other birds that are different from your first bird? If so, how many?

Part B: Design 2

1. Use a set of pattern blocks and the following directions to make the given design.
   - Start with a hexagon.
   - On each of the two topmost sides of the hexagon, attach a triangle.
   - On the bottom of the hexagon, attach a trapezoid.
   - Attach a square on the left and right sides of the trapezoid.

2. Compare your design to the design of one of your peers.
   a. Do your designs match? If not, how are they different?
   b. Is one more correct than the other based on the given directions? How do you know?

3. Can you use the same set of directions to make other designs that are different from your first? If so, how many?

Part C: Design 3

1. Use a set of pattern blocks and the following directions to make the given design.
    - Start with a hexagon. Position it so that it has two vertical sides.
    - On each of the 2 topmost non-vertical sides, attach a square so that the side of the square exactly matches the side of the hexagon.
    - Attach a triangle to the bottom of the hexagon so that a vertex of the triangle touches the vertex of the hexagon and one side of the triangle is horizontal.
    - Attach a trapezoid to the triangle so that the shortest parallel side of the trapezoid exactly matches the side of the triangle.

2. Compare your design to the design of one of your peers.
    a. Do your designs match? If not, how are they different?
    b. Is one more correct than the other based on the given directions? How do you know?

3. Can you use the same set of directions to make other designs that are different from your first? If so, how many?

Part D: Summary and Connections

1. Look back at the directions for the three designs. Did the language change from one set of directions to the next? If so, how?

2. Your designs may have differed from those of your peers. In such cases, what was it about the directions that allowed the designs to be different?

3. What does this exploration indicate about the importance of the precision and clarity of the language when communicating mathematically?

Exploration 1.3 Patterns of Common Traits

Purpose: Students use inductive reasoning to discover patterns of common traits and represent them in different ways.

Materials: Paper and pencil

Part A: Common Traits in Geometric Figures

In a **pattern of common traits**, the examples exhibit a common characteristic that can be found by analyzing their similarities and differences. The patterns can occur in a variety of situations. For instance, consider the following geometric figures:

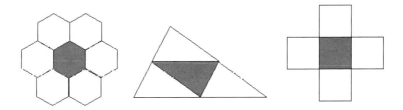

1. What common traits do the figures have? Write a description of any pattern you see.

2. a. If possible, use each shape to draw a geometric figure that satisfies the pattern you found in the previous problem.

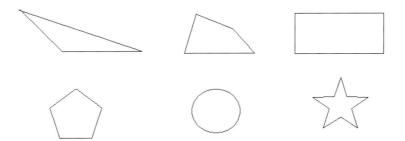

   b. If some of the shapes from part (a) did not satisfy the pattern you observed, can you explain why did they not work?
   c. If some of the shapes from part (a) did not satisfy the pattern you observed, can you change the pattern that you are considering so that they will work?

3. Can you observe a pattern of common traits in the figures that can be represented with numbers? If so, do so. If not, why not?

4. Can you observe a pattern of common traits in the figures that can be represented with a symbolic expression? If so, do so. If not, why not?

Part B: Common Traits in Multiplication Facts

Patterns of common traits can also occur in numerical situations. Use the following set of multiplication facts to answer each question.

$$3 \times 1 = 3 \qquad 4 \times 1 = 4 \qquad 1 \times 5 = 5$$
$$12 \times 1 = 12 \qquad 1 \times 11 = 11 \qquad 45 \times 1 = 45$$

1. What common traits do the products have? Write a description of any pattern you see.

2. a. Does your pattern hold true for other numbers? If so list five other multiplication facts that satisfy the same pattern.
   b. Does your pattern hold true for all numbers? If so, explain why. If not, provide a counterexample.

3. Can the pattern you see in the products be represented with a picture? If so, do so. If not, why not?

4. Can the pattern you see in the products be represented with a formula? If so, do so. If not, why not?

Part C: Summary and Connections

1. How did you use inductive reasoning to complete each part of this exploration?

2. What does this exploration suggest about using different representations to describe patterns?

Exploration 1.4 Growing Patterns

Purpose: Students use inductive reasoning to discover growing patterns and represent them in different ways.

Materials: Paper and pencil

Part A: A Pattern that Grows Larger

In a **growing pattern**, the examples continuously increase or decrease in a predictable fashion. These patterns can occur in a variety of situations. For instance, consider the following sequence of geometric figures:

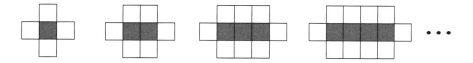

1. How do the figures grow from one figure to the next? Write a description of any pattern you see.

2. Based on the patterns you see, draw a picture of what you think the next three figures in the sequence will look like.

3. Not only can we represent the growing patterns verbally and visually, but we can also represent them numerically. Complete the table by counting the squares in each figure.

Figure in the Sequence	Number of Gray Squares	Number of White Squares	Total Number of Squares
1			
2			
3			
4			
5			
6			
7			

a. What growing patterns do you see in the number of
   i. Gray squares?   ii. White squares?   iii. Total squares?
b. How does the number of white squares grow compared to the number of gray squares? What about the total number of squares to the number of gray squares?

4. Many of the patterns in this sequence of figures can be represented symbolically. To do so, let n represent the position of the figure in the sequence.
   a. What else does n represent?
   b. Write a symbolic expression that represents the number of white squares in terms of n.
   c. Write a symbolic expression that represents the number of total squares in terms of n.
   d. Use your formulas from parts (b) and (c) to find the number of white squares and total squares in the figure that has
      i. 25 gray squares?   ii. 73 gray square?   iii. 1,010 gray squares?
   e. Why is a symbolic expression a powerful way to represent a growing pattern?

Part B: A Pattern that Grows Smaller

Growing patterns can also continuously decrease. For instance, consider the following list of numbers:

$$16, 8, 4, 2, \ldots$$

1. How do the numbers grow? Write a description of any pattern you see.

2. Based on the pattern you see, what are the next five numbers likely to be?

3. a. Can you draw a sequence of pictures that represents the pattern you see in the first four numbers in the list?
   b. What about the next four numbers in the list?

4. Can you write a symbolic expression to represent the pattern you see in the list of numbers? If so, what does the variable in your expression represent?

Part C: Summary and Connections

1. How did you use inductive reasoning to complete each part of this exploration?

2. Consider the list of numbers 10, 8, 6, 4, 2, ... Can you represent the next four numbers in the list with pictures? If so, do so. If not, why not?

Exploration 1.5 Venn Diagrams and Valid Arguments

Purpose: Students use Venn diagrams to determine the validity of arguments involving statements with quantifiers.

Materials: Paper and pencil

Part A: Venn Diagrams and Valid Arguments

Many of the arguments that we encounter use statements with **quantifiers**, which are words or phrases that imply a quantity. Common examples of quantifiers include *all*, *every*, *some*, *there exists*, or *none*. The relationships exhibited in many quantified statements, such as those of the form "All…are…" or "Some... are not ...," can be represented with a **Venn diagram**. In a Venn diagram, circles are used to represent the sets of objects in the quantified statements and dots are used to represent individual objects. The Venn diagrams for three different quantified statements are shown.

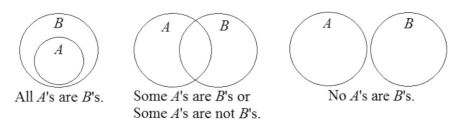

All A's are B's.   Some A's are B's or Some A's are not B's.   No A's are B's.

One way to determine whether arguments involving quantifiers make sense is to draw a Venn diagram so that each premise is true and then make a determination about the truthfulness of the conclusion. For example, consider the following argument:

> All cats are mammals.
> No lizard is a mammal.
> Therefore, no lizard is a cat.

To draw a Venn diagram that shows the first premise is true, make a circle for cats and place it entirely in another circle for mammals. For the second premise, draw a circle for lizards that is completely separate from the circle for mammals.

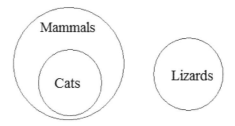

Because the circle for cats and the circle for lizards cannot have an animal in common, the conclusion of the argument is logically correct, and we say the argument is **valid**.

With some arguments, it is possible to draw the Venn diagram in two different ways and still have the premises true. In such cases, the conclusion is true in one instance but false in the other. Because we cannot determine which is correct, the argument must be logically incorrect, and we say that it is **invalid**.

1. Use a Venn diagram to determine the validity of each argument. Remember that individual objects can be represented with a dot and placed within the appropriate region of the diagram.

   a. Every chicken is a fowl.
      Every fowl can fly.
      Therefore, every chicken can fly.

   b. Some soccer players get good grades.
      Theresa is a soccer player.
      Therefore, Theresa gets good grades.

   c. All dolphins are intelligent animals.
      Flipper is an intelligent animal.
      Therefore, Flipper is a dolphin.

   d. No honest person cheats on tests.
      Joshua is not an honest person.
      Therefore, Joshua cheats on tests.

   e. Some Martians are green.
      Some Martians have three eyes.
      Therefore, some Martians are green and have three eyes.

2. Use a Venn diagram to find a conclusion that makes the given argument valid.

   a. All dogs are mammals.
      All mammals are warm blooded.
      Therefore, ...

   b. Every pizza is greasy.
      No greasy food is healthy.
      Therefore, ...

   c. All college students have cell phones.
      Roopa is a college student.
      Therefore, ...

   d. All third grade teachers teach math.
      Some people in this class will teach
         third grade.
      Therefore, ...

Part B: Summary and Connections

1. An argument is valid if the conclusion is true whenever the premises are true. How can you use truth tables to determine whether an argument is valid?

2. If an argument is logically valid, does that mean the argument is true?

Exploration 1.6 Lewis Carroll Logic Puzzles

Purpose: Students work on their reasoning ability by solving logic puzzles from Lewis Carroll's *Symbolic Logic*.

Materials: Paper and pencil

Part A: Lewis Carroll's Logic Puzzles

Lewis Carroll, the author of *Alice in Wonderland*, was the pen name of English mathematician and logician Charles Dodgson. Dodgson authored several other books in which he provides fanciful arguments that can be solved with deductive reasoning. Many of the arguments use *all* or *none* statements, which can be translated into conditional statements. For example, the statement "All gorillas are hairy" can be rewritten as "If the animal is a gorilla, then it is hairy." Likewise, the statement "No cats like dogs" can be rewritten as, "If it is a cat, then it does not like dogs." Once a statement is written in conditional form, it can be represented symbolically. The symbols can then be used to find a reasonable conclusion to the argument.

For example, suppose we want to find a logical conclusion to the argument:

> Everyone who is sane can do logic.
> No lunatics are fit to serve on a jury.
> None of your sons can do logic.
> Therefore, ...

To begin, pick a letter to represent each simple statement. Specifically, let s represent, "A person is sane," l represent, "A person can do logic," j represent, "A person can serve on a jury," and y represent, "A person is your son." Next, rewrite the argument verbally and symbolically as follows:

If a person is sane, then the person can do logic.	$s \to l$
If a person is not sane, then the person is not fit to serve on a jury.	$\sim s \to \sim j$
If a person is your son, then the person cannot do logic.	$y \to \sim l$
Therefore, ...	$\therefore$

Using the symbols and a contrapositive, we can rearrange the statements to get:

If a person is your son, then the person cannot do logic.	$y \to \sim l$
If a person cannot do logic, then the person is not sane.	$\sim l \to \sim s$
If a person is not sane, then the person is not fit to serve on a jury.	$\sim s \to \sim j$
Therefore, ...	$\therefore$

From this, we conclude that $y \rightarrow \sim j$, or "If a person is your son, then the person is not fit to serve on a jury." Translating the statement back to the language of the original argument, we have, "No son of yours is fit to serve on a jury."

The following are several of Carroll's puzzles. Use conditional statements and symbolic logic to find a correct conclusion for each one.

1. My saucepans are the only things that I have that are made of tin.
   I find all your presents very useful.
   None of my saucepans is of the slightest use.
   Therefore, ...

2. No terriers wander among the signs of the zodiac.
   Nothing that does not wander among the signs of the zodiac is a comet.
   Nothing but a terrier has a curly tail.
   Therefore, ...

3. Nobody who really appreciates Beethoven fails to keep silent while the
      Moonlight Sonata is being played.
   Guinea-pigs are hopelessly ignorant of music.
   No one who is hopelessly ignorant of music ever keeps silent while the
      Moonlight Sonata is being played.
   Therefore, ...

4. No birds, except ostriches, are nine feet high.
   There are no birds in this aviary that belong to anyone but *me*.
   No ostrich lives on mince-pies.
   I have no birds less than nine-feet high.
   Therefore, ...

5. No kitten that loves fish is unteachable.
   No kitten without a tail will play with a gorilla.
   Kittens with whiskers always love fish.
   No teachable kitten has green eyes.
   No kittens have tails unless they have whiskers.
   Therefore, ...

Part B: Summary and Connections

1. What argument form was used in each argument of this exploration?

2. How might activities like this be useful in teaching students mathematics? What about writing skills and reading comprehension?

Exploration 1.7 The Sum of an Even and an Odd Number

Purpose: Students use inductive and deductive reasoning to explore and represent facts about the sum of an even and an odd number.

Materials: Square tiles or connecting cubes

Part A: Inductive Reasoning and the Sum of an Even and an Odd Number

1. Fekadu claims that the sum of an even and an odd number is always odd. How can you use inductive reasoning to test his conjecture?

2. After testing Fekadu's claim with several examples, do you believe Fekadu is correct? Why or why not?

3. Suppose you tested Fekadu's claim with 100 pairs of an even and an odd number, and each time you obtained an odd sum. Do the 100 pairs guarantee the sum of any even number and any odd number is always odd? What about 500 pairs? 1,000 pairs?

Part B: Deductive Reasoning and the Sum of an Even and an Odd Number

In Part A, you used pairs of numbers to test whether the sum of an even and an odd number is odd. However, is this always true? One way to verify the conjecture is to offer a visual proof.

1. Discuss with a group of your peers what makes a number even and what makes a number odd. Decide how you might use your ideas to represent an even and an odd number with square tiles or connecting cubes. Draw a picture of how you would represent each of the following numbers in this way.
    a. 5   b. 8   c. 10   d. 13

2. Adapt your representation to show how you might represent the sum of an even and an odd number. Draw a picture of how you would represent each sum in this way. How do your pictures illustrate that the sum of an even and an odd number is odd?
    a. 2 + 3   b. 4 + 7   c. 6 + 1   d. 8 + 3

3. Adapt your representation to show how you might represent the sum of *any* even with any odd number. What must you do to communicate that your representation is now for any even or any odd? Draw each of your representations.

4. Use your pictures from the previous question to help you write an argument that explains why the sum of an even and an odd number is always odd.

5. Why is the reasoning used in Part B deductive rather than inductive?

Part C: Summary and Connections

1. Use your pictures from Part B to help you write symbolic expressions that represent any even or odd number. Use your symbolic expressions to write an argument that shows the sum of an even and an odd number is odd.

2. What does this exploration reveal about the use of pictures in deductive reasoning?

3. Children often explore the sums of even and odd numbers in the elementary grades.
    a. Do you think children in the elementary grades could make sense of the visual proof discussed in this exploration?
    b. How might visual proofs help you to include more deductive reasoning in your classroom?

Exploration 1.8 The Sum of Three Consecutive Numbers

Purpose: Students use inductive and deductive reasoning to explore and represent facts about the sum of three consecutive counting numbers.

Materials: Paper and pencil

Part A: Inductive Reasoning and the Sum of Three Consecutive Numbers

1. In this exploration, we consider the sum of three consecutive numbers. What does it mean for three numbers to be consecutive?

2. Make a list of ten different sets of three consecutive numbers. Find the sum of the numbers in each set. Write a description of any pattern you see in the sums.

3. Is your conjecture always true? Test it using other sets of consecutive numbers.

Part B: Deductive Reasoning and the Sum of Three Consecutive Numbers

In Part A, you may have noticed that the sum of every one of your sets of three consecutive numbers was divisible by 3. This is always the case, and the next several questions guide you through a symbolic proof of this fact.

1. Discuss with a group of your peers what it means for a number to be divisible by 3.

2. In order to prove that the sum of three consecutive numbers is divisible by three, the sum of the numbers must first be represented symbolically. If n represents the smallest of the numbers, how can you represent the next two consecutive numbers? How can you then represent the sum of all three?

3. Use a few algebraic manipulations to write an argument that shows the sum is divisible by 3. Be sure to explain each step of your argument.

4. Why is the reasoning used in Part B deductive rather than inductive?

Part C: Summary and Connections

1. Is it possible to represent the sum of three consecutive numbers with a picture? If so, use your picture to write a different argument that shows the sum of three consecutive numbers is divisible by 3.

2. What does this exploration reveal about the use of symbolic expressions in deductive reasoning?

Exploration 1.9 Cutting a Pie

Purpose: Students solve a problem by answering specific questions asked in each step of Polya's problem-solving method.

Materials: Paper and pencil

Part A: Understanding the Problem

Solve the following problem by answering the questions in each step of Polya's problem-solving method.

A pie can be cut into four pieces using two straight cuts and into seven pieces using three straight cuts. What is the largest number of pieces that a pie can be cut into using eight straight cuts?

1. What is the goal of the problem?

2. a. What information is given in the problem?
   b. Is there any information that is needed but not given?

Part B: Devising a Plan

1. What strategies might be useful in solving this problem?

2. How does your common sense, mathematical knowledge, or previous experience help you devise a plan for solving this problem?

Part C: Carrying Out the Plan

Use your plan from Part B to solve the problem. If necessary, revise your plan.

Part D: Looking Back

1. Does your solution answer the question? How do you know it is correct?

2. Do you see another way to solve the problem? If so, which way is more efficient?

Part E: Summary and Connections

1. What step of Polya's method was most useful in solving this problem? Least useful?

2. Do you think teaching students to use Polya's method will help them to become better problem solvers? Why or why not?

Exploration 1.10 Seating at a High School Graduation

Purpose: Students solve a problem by answering specific questions asked in each step of Polya's problem-solving method.

Materials: Paper and pencil

Part A: Understanding the Problem

Solve the following problem by answering the questions in each step of Polya's problem-solving method.

Thien is at her high school graduation. The chairs for the graduates are divided into two equal sections separated by a center aisle. Thien is sitting in the sixth row on the right side and there are eight rows of chairs behind her. There are six students to her right and five to her left in her row. If a graduate is sitting in each chair and all graduating students are in attendance, how many students are graduating?

1. What is the goal of the problem?

2. a. What information is given in the problem?
   b. Is there any information that is needed but not given?

Part B: Devising a Plan

1. What strategies might be useful in solving this problem?

2. How does your common sense, mathematical knowledge, or previous experience help you devise a plan for solving this problem?

Part C: Carrying Out the Plan

Use your plan from Part B to solve the problem. If necessary, revise your plan.

Part D: Looking Back

1. Does your solution answer the question? How do you know it is correct?

2. Do you see another way to solve the problem? If so, which way is more efficient?

Part E: Summary and Connections

Write a summary of how useful Polya's method was to you in solving this problem. Specifically, address which step was most useful and which step was least useful.

Exploration 1.11 How Much Do School Supplies Cost?

Purpose: Students use problem-solving skills to solve a real-world problem involving a budget for an elementary classroom.

Materials: Paper and pencil

Part A: Buying School Supplies on a Budget

You were recently hired to teach third grade at a local elementary school. As a new faculty member, your principal has given you an allowance of $750 to buy supplies for your classroom. Make a list of the supplies you would purchase with this money. Use exact costs and assume you have 22 students in your class.

Part B: Summary and Connections

1. a. What information did you need to complete this problem?
   b. What resources did you use to find this information?

2. What assumptions did you make about your classroom or your students to complete this problem? Do you think your assumptions were realistic and accurate? Why or why not?

3. What steps of Polya's problem-solving method did you use either formally or informally to solve this problem?

4. This exploration suggests only one possible way that you might use problem solving in your future career as a teacher. What are some others?

Exploration 1.12 Humans and Horses

Purpose: Students examine how to use a variety of strategies to solve one problem.

Materials: Paper and pencil

Part A: Humans and Horses

The following problem can be solved using a variety of strategies. Before we look at some specific strategies, try it on your own. After you have completed the problem, write a short description of the strategy you used to solve the problem.

> In a crowd of humans and horses, there are 48 legs and 18 heads. How many horses are in the crowd?

Part B: Drawing a Picture

One way to solve the problem is to draw a picture. To do so, let an oval represent a head and then draw segments off the oval for the legs as shown in the following figure. Solve the problem using simple drawings like this. After you are finished, write a sentence or two describing how you used the pictures to solve the problem.

Human   Horse

Part C: Guess and Check

Another way to solve the problem is to make and check a number of guesses. It may be helpful to use and extend a table like the following to keep track of your guesses as you make them. One guess has been made. After you have your answer, write a sentence or two that summarizes how you solved the problem in this way.

Number of Humans	Number of Horses	Total Number of Legs
8	10	8(2) + 10(4) = 16 + 40 = 56

Part D: Write an Equation

A third approach is to write an equation. In this problem, there are two unknowns, the number of humans, n, and the number of horses, h. Use the information in the problem to write an equation or equations to solve the problem. After you have your answer, write a sentence or two that summarizes how you solved the problem in this way.

Part E: Summary and Connections

1. Of the different strategies used to solve this problem, which do you think was the easiest to use? Why?

2. What does this exploration reveal about solving mathematical problems? Does it influence how you might teach problem solving in your future career as a teacher?

Chapter 2
Sets

Sets and set operations play an important role in elementary mathematics. Not only do they provide the foundation for numeration and numerical operations, but they are also crucial to counting and sorting. **Counting** is the method we use to identify and name quantities that are difficult to perceive in other ways. To count correctly, children must first master the idea of a one-to-one correspondence, which is a fundamental way to compare two sets. Only after they have mastered this idea and are counting correctly will they begin to develop the idea of number. Children also use sets and set operations to sort. **Sorting**, or **categorizing**, is the process of placing objects or events into sets based on their similarities and differences. Sorting skills are important to children both cognitively and mathematically. Cognitively, children use sorting to develop language, organize memories, and relate new experiences to old. Mathematically, children use sorting skills to recognize patterns, identify computational properties, classify shapes, and sift through data.

Many of the explorations in this chapter use attribute blocks to investigate the ideas associated with sets, set relationships, and set operations. Others connect the content of sets to the elementary classroom. Specifically,

- Explorations 2.1 - 2.3 introduce attribute blocks and use them to investigate set relationships and solve problems.
- Exploration 2.4 considers the different types of counting often used in the elementary grades.
- Explorations 2.5 - 2.6 use attribute blocks to explore set operations and to solve sorting problems.

Exploration 2.1 An Introduction to Attribute Blocks

Purpose: Students become familiar with attribute blocks by examining the different subsets contained within a standard set of 60 attribute blocks.

Materials: A set of 60 attribute blocks.

Part A: Attribute Blocks

A set of 60 attribute blocks has four attributes: shape, color, size, and thickness. The specific attributes and the way they are denoted are as follows:

Shape	Color	Size	Thickness
Circle = ○	Red = R	Large = L	Thick = T
Hexagon = ⬡	Blue = B	Small = S	Thin = Th
Rectangle = ▭	Yellow = Y		
Square = □			
Triangle = △			

Each symbol is used to represent the set of all shapes with the given attributes. For instance, B represents the set of all blue attribute blocks, L represents the set of all large attribute blocks, and $Th○$ represents the set of all thin circles.

1. Complete the table by giving a verbal description of each set and its cardinal number.

Set	Verbal Description	Cardinal Number
□		
Y		
S		
$B△$		
LTh		
$T⬡$		
TLR		
$SY○$		
$TLB▭$		
$ThSR△$		

2. There are twelve blocks in the set of rectangles. Use the roster method to list them.

3. Use a set of 60 attribute blocks to answer each question.
    a. How many triangles are
        i. Blue?   ii. Small?   iii. Thin?

    b. How many red blocks are
        i. Hexagons?   ii. Large?   iii. Thick?

    c. How many small blocks are
        i. Rectangles?   ii. Yellow?   iii. Thin?

    d. How many thick blocks are
        i. Circles?   ii. Red?   iii. Large?

4. Use a set of 60 attribute blocks to answer each question.
    a. How many blocks of each shape will be each
        i. Color?   ii. Size?   iii. Thickness?

    b. How many blocks of each color will be each
        i. Shape?   ii. Size?   iii. Thickness?

    c. How many blocks of each size will be each
        i. Shape?   ii. Color?   iii. Thickness?

    d. How many blocks of each thickness will be each
        i. Shape?   ii. Color?   iii. Size?

    e. How many blocks will have any one particular combination of all four attributes?

Part B: Summary and Connections

1. How many attributes do any two blocks have in common if they differ by one attribute? By two? By three? By four?

2. Attribute blocks are not the only manipulatives that can be used to sort. Search the Internet to find other materials and ideas that can be used to sort by attributes.

Exploration 2.2 Difference Puzzles

Purpose: Students become familiar with attribute blocks by using them to solve difference puzzles.

Materials: A set of 60 attribute blocks.

Part A. Difference Trains

A **difference train** is a sequence of attribute blocks in which each successive block differs from the previous block by a set number of attributes. For instance, the following is a difference train in which each block differs from the previous block by exactly two attributes.

1. Create a difference train of six attribute blocks where each successive block differs from the previous block by
    a. Three attributes.   b. Four attributes.

2. Create a difference train of six red attribute blocks where each successive block differs from the previous block by one attribute.

3. Create a difference train of six thick attribute blocks where each successive block differs from the previous block by two attributes.

4. Is it possible to create a difference train that uses all 60 attribute blocks so that each successive block differs from the previous block by three attributes? Describe your strategy for completing this task if it is possible.

Part B: Arrow Diagrams

Complete the puzzles on the Arrow Diagrams sheet. Start by placing an attribute block in any square. Next, place attribute blocks in the rest of the squares so that the number of differences between the attributes of any two blocks matches the number of arrows between the corresponding squares.

Part C: Difference Grids

Use the Difference Grids sheet to complete the following puzzles.

1. The first puzzle uses a 3 × 3 grid. Using only three small shapes, but any color or thickness, complete the grid by placing blocks in the squares so that no two blocks of the same shape or color occupy the same row or column.

2. The second puzzle is a variation of the grid puzzle. Using only small attribute blocks, place blocks in the grid so that adjacent blocks differ by the number of attributes shown in the small squares.

Part D: Summary and Connections

1. Describe how you used mathematical reasoning to complete the puzzles in this exploration.

2. With one or two of your classmates, design an activity that teaches children to sort attribute blocks first by 1 difference, then by 2, and then by 3.

Arrow Diagrams

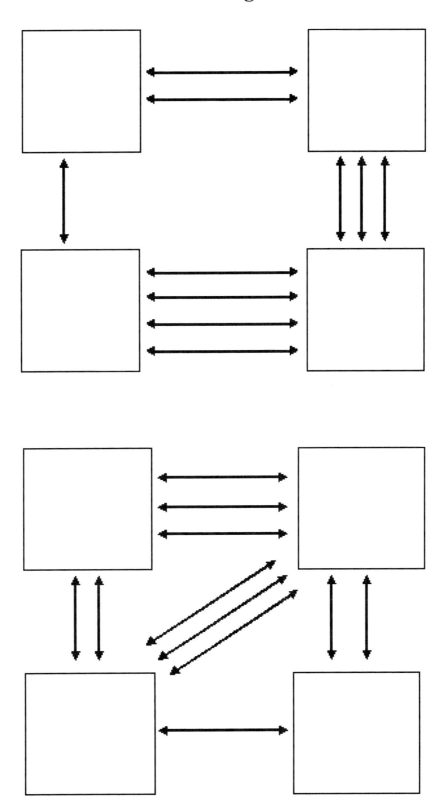

Difference Grids

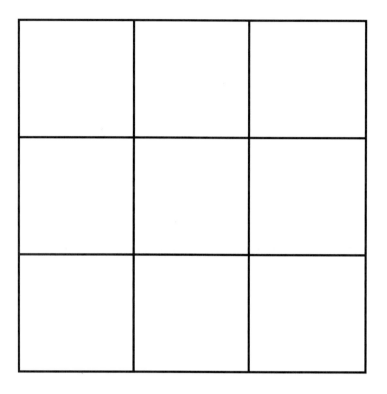

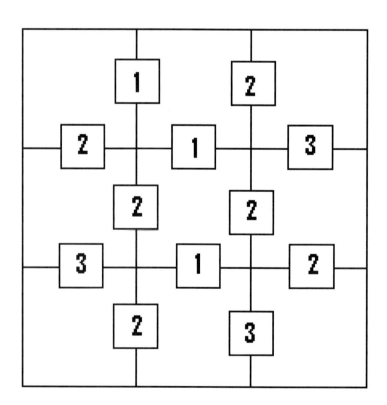

Exploration 2.3 Subsets, Equal Sets, and Equivalent Sets

Purpose: Students examine subsets, equal sets, and equivalent sets by comparing different sets of attribute blocks.

Materials: A set of 60 attribute blocks.

Part A: Subsets

Sets can relate to one another in a number of ways. One such relationship is the subset relationship, or that of containment. Set A is said to be a **subset** of set B, written $A \subseteq B$, if and only if, every element of A is also an element of B. If B has an element that is not in A, then A is a **proper subset** of B, written $A \subset B$. Use a set of 60 attribute blocks to answer the following questions about subsets.

1. Give a verbal interpretation of each statement and determine whether it is true or false.
    a. $R\triangle \subseteq R$.
    b. $R \subseteq TR$.
    c. $B \subseteq L$.
    d. $Y\square \subseteq \square$.
    e. $LTY \subseteq TY$.
    f. $YTS\bigcirc \subseteq S$.

2. a. List the subsets of the set $TB\square$. Circle all those that are proper subsets.
   b. List of the subsets of the set $LTh\triangle$. Circle all those that are proper subsets.
   c. In each of the previous parts, what was the only subset that was not a proper subset? Is this always true? Use the definitions of subsets and proper subsets to explain.

3. Consider the set Y and the set ThY.
    a. Which set is a subset of the other?
    b. Draw a Venn diagram of the two sets and then place the blocks from each set into the appropriate region of the diagram. How does your diagram illustrate the subset relationship between the sets?

Part B: Equal and Equivalent Sets

Sets can also be equal or equivalent. Two sets are **equal** ($A = B$) if and only if they contain exactly the same elements. Two set are **equivalent** ($A \sim B$) if and only if there is a one-to-one correspondence between them. A **one-to-one correspondence** is a pairing of the elements of A with the elements of B so that each element of A corresponds to exactly one element of B, and vice versa. Intuitively, two sets are equal if they contain exactly the same elements and equivalent if they contain the same number of elements.

1. If $A = \{1, 2, 3\}$, what elements must be in a set B for $A = B$?

2. Consider a set of 60 attribute blocks. Are any subsets of the attribute blocks equal? Explain.

3. Consider a set of 60 attribute blocks. The set of thin squares (*Th*□) is equivalent to the set of thick squares (*T*□) because a one-to-one correspondence can be made between them.
    a. Draw a diagram that illustrates one such one-to-one correspondence. How did you illustrate the paring of the elements?
    b. Is this the only one-to-one correspondence between *Th*□ and *T*□? If not, draw a different one from the one you drew in part (a).
    c. Does it matter which one-to-one correspondence is used to show the sets are equivalent? Why or why not?

3. Is $BO \sim RO$? Make a one-to-one correspondence to explain how you know.

4. Is $L\triangle \sim B\square$? Make a one-to-one correspondence to explain how you know.

5. List two other sets that are equivalent to
    a. *L* b. *B* c. *O* d. *ST* e. *ThY* f. *L*△

Part C: Summary and Connections

1. a. If A is a proper subset of B, must A be a subset of B?
   b. If A is a subset of B, must A be a proper subset of B?

2. Can you think of a situation in which finding or counting subsets would be a useful strategy for solving a problem? If so, write a story problem that can be solved in this way.

3. a. If A is equal to B, must A be equivalent to B? Explain.
   b. If A is equivalent to B, must A be equal to B? Explain.

Exploration 2.4 Counting in the Elementary Classroom

Purpose: Students consider different ways of counting that are commonly used in the elementary classroom.

Materials: Colored chips or other counters

Part A: Counting On

When learning to count, children must master four skills. They must learn to

- Recite the number word sequence.
- Make a one-to-one correspondence between the set of number words and a set of objects to be counted.
- Keep track of the objects they have counted.
- Recognize that the last word used in the number sequence gives the number of objects in the set.

After children have mastered these skills, they can learn other counting strategies such as counting on. When counting on, children begin at a certain place in the number word sequence and then count on from that number. For example, to count on from five, a child would start at five and count "six, seven, eight...."

1. How does the use of the number word sequence change when a child counts on?

2. Is a one-to-one correspondence still used when counting on? If so, how does its use change? Draw a diagram that illustrates how a one-to-one correspondence is used to count on from five.

Part B: Counting Back

Children also learn to count back. When counting back, children begin at a certain place in the number word sequence and then count backwards. For example, to count back from ten, a child would start at ten and then count, "nine, eight, seven...."

1. How does the use of the number word sequence change when a child counts back?

2. Is a one-to-one correspondence still used when counting back? If so, how does its use change? Draw a diagram that illustrates how a one-to-one correspondence is used to count back from ten.

3. What might be different about the way children keep track of the objects they have counted when they count back compared to regular counting or counting on?

Part C: Skip Counting

Another counting strategy is skip counting. When skip counting, children begin at a certain place in the number word sequence and then count on by a certain number such as two, five, or ten. For example, to skip count by twos a child would start at two and then count, "four, six, eight…."

1. How does the use of the number word sequence change when a child skip counts?

2. Is a one-to-one correspondence still used when skip counting? If so, how does its use change? Draw a diagram that illustrates how a one-to-one correspondence is used to skip count by twos.

3. What might be different about the way children keep track of the objects they have counted when skip counting compared to regular counting or counting on?

4. Repeat questions 1 – 3 only this time use skip counting by fives.

Part D: Summary and Connections

1. How might counting on, counting back, and skip counting be used to teach whole-number operations?

2. Use each of the numbers two through ten to skip count several numbers. Which numbers are easiest to use? Most difficult to use? Why do you think this is the case?

Exploration 2.5 Attribute Blocks and Set Operations

Purpose: Students represent and compute set operations using attribute blocks.

Materials: A set of 60 attribute blocks.

Part A: The Union

The **union** of two sets A and B, written $A \cup B$, is the set of all elements in A or in B or in both. In the definition, *or* is a key word that indicates an object is in the union if it is in one set or the other, or possibly in both sets. Consider a set of 60 attribute blocks. Write each union symbolically, then determine how many elements it has.

1. Red blocks or thick blocks
2. Yellow blocks or hexagons
3. Large blocks or thin blocks or squares
4. Yellow squares or blue hexagons
5. Large hexagons or thin triangles.
6. Large blocks or thin blocks

Part B: The Intersection

The **intersection** of two sets A and B, written $A \cap B$, is the set of all elements common to A and B. In the definition, *and* is a key word that indicates that an object is in the intersection only if it occurs in both sets.

1. Using a set of 60 attribute blocks, write each intersection symbolically. Then describe the intersection in terms of a single set and determine **how many** blocks are in it.
    a. Yellow blocks and large blocks
    b. Thin small blocks and red hexagons
    c. Small blocks and thin blocks
    d. Small rectangles and blue squares
    e. Large blocks and blue blocks and rectangular blocks

2. Use a set of 60 attribute blocks to answer each question.
    a. Find two sets with an intersection of 6 elements.
    b. Find two sets with an intersection of 4 elements.
    c. Find three sets with an intersection of 5 elements.
    d. Two sets are **disjoint** if the intersection of the sets is empty. List three pairs of disjoint sets.

Part C: The Set Difference

The **set difference** of set B from set A, written $A - B$, is the set of all elements in A that are not in B. The word *from* indicates an order to this operation; we look at what elements are left in A after any elements common to A and B have been removed.

1. Use a set of 60 attribute blocks. First give another interpretation for each set difference. For instance, the set of yellow blocks from the large blocks can also be described as the set of all large blocks that are not yellow. Then, write each set difference symbolically and determine **how many** blocks are in it.
   a. Red blocks from thick blocks
   b. Small thin blocks from small blocks
   c. Blue blocks from all blocks
   d. Yellow squares from all squares
   e. Large red hexagons from large red blocks.

2. Consider the sets $Y - T$ and $T - Y$. Are the two sets equal? Equivalent? Use attribute blocks to write a short argument that justifies your thinking.

Part D: The Complement

The **complement** of a set A, written $\overline{A}$, is the set of all elements in the universal set U that are not in A.

1. Use a set of 60 attribute blocks. Write the complement of each set symbolically and determine how many blocks are in it.
   a. Blue blocks
   b. Small thin blocks
   c. Large triangles or small rectangles
   d. Yellow squares or blue hexagons
   e. Large hexagons or thin triangles.

2. The complement of a set A can also be written as $U - A$.
   a. Why does it make sense to write the complement as a set difference?
   b. Use a set difference to represent each complement in question 1.

Part E: Summary and Connections

1. How does the word *and* differ from the word *or* when computing set operations?

2. With several of your classmates, design a short activity for elementary students that uses attribute blocks and has questions that can be solved with a
   a. Union.
   b. Intersection.
   c. Set Difference.
   d. Complement.

Exploration 2.6 Sorting by Two and Three Attributes

Purpose: Students use Venn diagrams to sort attribute blocks by two and three attributes.

Materials: A set of 60 attribute blocks and three pieces of string, each about 2 feet

Part A: Sorting by Two Attributes

1. Form a Venn diagram on your desk by arranging two strings in overlapping circles. Let one circle represent the set of blue blocks and the other represent the set of large blocks. The circles divide your desk into four regions as indicated by the Roman numerals in the following diagram. Sort a set of 60 attribute blocks by placing them into the appropriate regions on your desk. After you are finished, complete the table.

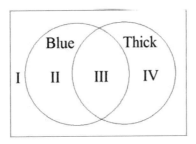

Region	Verbal Description	Set Notation	Number of Blocks
I			
II	Blue blocks that are not thick	$B - T$	10
III			
IV			

2. Repeat the previous exercise, but this time let one circle represent the set of thin blocks and the other represent the set of triangles.

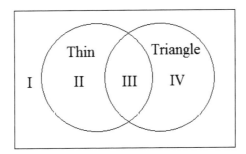

Region	Verbal Description	Set Notation	Number of Blocks
I			
II			
III			
IV			

Part B: Sorting by Three Attributes

1. Form a Venn diagram on your desk by arranging three strings into overlapping circles. Let one circle represent the set of yellow attribute blocks, another represent the set of thin blocks, and the third represent the set of large blocks. The circles divide your desk into eight distinct regions as indicated by the Roman numerals in the following diagram. Sort a set of 60 attribute blocks by placing them into the appropriate regions on your desk. After you are finished, complete the table.

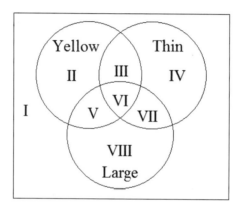

Region	Verbal Description	Set Notation	Number of Blocks
I			
II			
III			
IV			
V			
VI			
VII			
VIII			

2. Repeat the previous exercise, but this time let one circle represent the set of squares, another represent the set of red blocks, and the third represent the set of thick blocks.

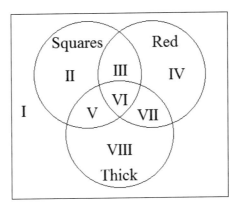

Region	Verbal Description	Set Notation	Number of Blocks
I			
II			
III			
IV			
V			
VI			
VII			
VIII			

Part C: Summary and Connections

1. What was the universal set for each problem in this exploration?

2. Describe the process you used to complete each of the Venn diagrams? What role did mathematical reasoning play in this process?

3. What role did set operations play in completing this exploration? What does this suggest about how students will use set operations in the elementary classroom?

Chapter 3
Numbers and Numeration

Once children have learned to count, they can use it to learn about numbers, numeration, and numerical operations. It is essential that children master base-ten numeration and numerical operations, as they are the fundamental building blocks for almost all of mathematics. Chapter 3 encompasses two important topics from elementary mathematics, base-ten numeration and number sense. In the elementary grades, children learn to write and understand decimal numerals. In doing so, they learn about place value, groupings, and exchanges. These ideas are difficult for some students to master, but it is essential that they do so. They are not only important to writing numerals, but also to computing with large numbers.

As children learn numeration, they also develop number sense. This entails learning the values of numerals and their relative size to one another. It also involves representing and using numbers in different ways. If children can be flexible in the way they think about numbers, then they can better apply what they know by choosing the representation that is best suited to their needs. In the early grades, this primarily involves making exchanges between base-ten place values.

Many of the explorations in this chapter will help you gain a deeper understanding of the characteristics of the decimal system by having you work in a variety of numeration systems. Other explorations are designed to help you develop your number sense using base-ten blocks, which are a common manipulative used in the elementary grades. Specifically,

- Explorations 3.1 - 3.3 explore the concepts of numeration by considering the characteristics of different numeration systems.
- Explorations 3.4 - 3.5 use base-ten blocks to understand decimal numerals and develop number sense.
- Exploration 3.6 considers numerals in base five and base eight.

Exploration 3.1 Exploring Mayan Numeration

Purpose: Students discover features of the Mayan numeration system by comparing it to the decimal system.

Materials: Paper and pencil

Part A: One- and Two-Digit Numbers

Many features of Mayan numeration are similar to those of the decimal system. Consequently, Mayan numerals can be explored by comparing them to decimal numerals. Consider the following table that shows how to write the same number in both systems. Compare the Mayan numeral to its corresponding decimal numeral. Use the patterns you see to answer the questions.

Decimal Numeral	4	8	10	13	17
Mayan Numeral	••••	••• (over bar)	=	••• (over two bars)	•• (over three bars)

1. How many symbols are used to write these Mayan numerals? Can you determine the value of each symbol?

2. Use the patterns you see to write the rest of the Mayan numerals for 1 to 19.

3. Use your observations with these numerals to write the Mayan numerals for 27 and 30.

Part B: Two- and Three-Digit Numbers

It is also possible to write the Mayan numerals that correspond to two- and three-digit decimal numerals. Compare the Mayan numerals to their corresponding decimal numerals in the following table. Use the patterns you see to answer the questions.

Decimal Numeral	20	28	43	102	140	267
Mayan Numeral	• / shell	• / ••• (over bar)	•• / •••	— / ••	•• / shell	••• (over bar) / •• (over bar)

1. What new symbol is introduced with these Mayan numerals? What is its value?

2. What changes occur in the Mayan numerals at 20? Can you explain the change?

3. Use your observations to write the Mayan numerals for 135 and 213.

44 Chapter 3 Numbers and Numeration

Part C: Larger Numbers

It is also possible to write the Mayan numerals for even larger values. Compare the Mayan numerals to their corresponding decimal numerals. Use the patterns you see to answer the questions.

Decimal Numeral	360	421	729	1,836	4,687
Mayan Numeral	• ⊘ ⊘	• ••• •	•• ⊘ ••••	― • ≡	••• ≡ ⊘ ••

1. a. These Mayan numerals indicate that the system must be positional. Why?
   b. What are the values of the positions?
   c. What changes occur at the third place value? Why do you think this occurred?
   d. Based on your observations, what do you think the fourth, fifth, and sixth place values will be?

2. Use your observations to write the Mayan numerals for 3,600 and 7,200.

Part D: Summary and Connections

1. How are the Mayan and decimal systems similar? How do they differ?

2. Is it possible to represent any whole number with Mayan numerals? Why or why not?

3. Suppose one of your classmates missed today's lab. Write a letter to this classmate explaining the characteristics of the Mayan system and how it works.

Exploration 3.2 Efficiency in Different Numeration Systems

Purpose: Students compare numerals written in different numeration systems to determine which system is the most efficient.

Materials: Paper and pencil

Part A: Multiples of Five

The power of a numeration system is partly determined by its efficiency. In general, a numeration system is more efficient if it requires fewer symbols to write numerals. Review how to write numerals in the Egyptian, Roman, Babylonian, and Mayan numeration systems. Next, complete the following table by writing the given multiples of five in the different numeration systems. After you are finished, compare how many total symbols or digits are needed to write each numeral. Which system appears to be the most efficient for writing multiples of five?

Multiple of Five	Egyptian	Roman	Babylonian	Mayan	Decimal
Five					
Fifteen					
Twenty-five					
Thirty-five					

Part B: Powers of Ten

Complete the following table by writing the given powers of ten in the given numeration systems. Which system appears to be the most efficient for writing powers of ten?

Powers of Ten	Egyptian	Roman	Babylonian	Mayan	Decimal
Ten					
One hundred					
One thousand					
Ten thousand					

© 2014 Cengage Learning. All Rights Reserved. May not be scanned, copied or duplicated, or posted to a publicly accessible website, in whole or in part.

Part C: Multiples of Nine

Complete the following table by writing the multiples of nine in the given numeration systems. Which system appears to be the most efficient for writing multiples of nine?

Multiple of Nine	Egyptian	Roman	Babylonian	Mayan	Decimal
Nine					
Eighteen					
Twenty-seven					
Thirty-six					

Part D: The Most Efficient Numeration System

Compare your answers from the last three parts and rank the numeration systems from most efficient to least efficient. If you cannot decide between two systems, select other numbers and write them in those systems. When you are finished, write a short paragraph that justifies why you ranked the numeration systems the way that you did.

Part E: Summary and Connections

1. The decimal system is the most common numeration system used throughout the world. Why do you think the decimal system was chosen over other historical systems?

2. Another factor that determines the power of a numeration system is how easy it is to compute in that system. Think about what it might be like to add, subtract, multiply, and divide in the Egyptian, Roman, Babylonian, and Mayan systems. In which system do you think computations would be easiest to perform and why?

Exploration 3.3 A Trip to Hexaminia

Purpose: Students learn a new numeration system to reinforce the concept of place value and to understand the regrouping process.

Materials: Paper and pencil

Part A: Understanding Hexaminian

You and one of your classmates have agreed to study abroad in the distant island of Hexaminia. As a part of your studies, you must learn how to use the Hexaminian numeration system. You learn that the Hexaminians can write every whole number using different combinations of only six symbols. To help you understand the value of each symbol and the way the numeration system works, your instructor gives you a table containing the first ten numerals and their names. Because the Hexaminians are unfamiliar with the decimal system, your instructor shows you the equivalent value of each numeral using single pebbles.

Hexaminian Numeral	Name	Equivalent Number of Pebbles
○	Zero	
|	Uni	●
V	Bi	● ●
△	Tri	● ● ●
□	Tetra	● ● ● ●
⬠	Penta	● ● ● ● ●
|○	Uni-hen	● ● ● ● ● ●
||	Uni-hen uni	● ● ● ● ● ● ●
|V	Uni-hen bi	● ● ● ● ● ● ● ●
|△	Uni-hen tri	● ● ● ● ● ● ● ● ●
|□	Uni-hen tetra	● ● ● ● ● ● ● ● ● ●

1. a. Based on patterns you see in the table, is the Hexaminian system a grouping system? How do you know?
   b. If it is a grouping system, what is the base?

2. Based on patterns you see in the table, does the Hexaminian system have place value? How do you know?

3. Use your observations to complete the following table using Hexaminian numerals and their names.

Collection of Pebbles	Hexaminian Numeral	Hexaminian Numeral Name
(11 pebbles)		
(10 pebbles)		
(13 pebbles)		
(14 pebbles)		
(15 pebbles)		
(16 pebbles)		
(17 pebbles)		
(18 pebbles)		

Part B: Larger Hexaminian Numerals

Your instructor tells you that it is also possible to write much larger numbers using Hexaminian numeration. To illustrate this, he gives you the following table of Hexaminian numerals and names.

Hexaminian Numeral	Numeral Name
V △	Bi-hen tri
I ○ ⬠	Uni-hundred zero-hen penta, or Uni hundred penta
△ ⬠ □	Tri-hundred penta-hen tetra
V ○ ⬠ ⬠	Bi-housand, zero hundred penta-hen penta, or bi-housand penta-hen penta
V I □ △	Bi-housand, uni hundred tetra-hen tri

1. Using your observations from the table, give the name of each Hexaminian numeral.

    a. □ ⬠   b. I V V   c. △ △ ○   d. V I I V

2. Using your observations from the table, give the numeral for each Hexaminian numeral name.
   a. Tri-hen penta
   b. Penta-hen
   c. Uni-hundred bi-hen tetra
   d. Bi-housand, tri-hundred penta-hen uni

3. a. Represent the number of stars with a Hexaminian numeral and then give the name of the numeral.

   b. Suppose there were tetra stars fewer than those shown above. Represent this amount with a Hexaminian numeral and then give its name.
   c. Suppose there were penta stars more than those drawn above. Represent this amount with a Hexaminian numeral and then give its name.

4. Using your observations from the previous table and the one in Part A, draw the number of stars equivalent to
   a. V⬠
   b. | ◯ V

5. If you are counting in the Hexaminian system, what Hexaminian numeral comes after each of these?
   a. ☐V
   b. | ◯ ⬠
   c. ⬠ ◯ ⬠
   d. Tri-hen tetra
   e. Tetra-hundred penta-hen penta

6. If you are counting in the Hexaminian system, what Hexaminian numeral comes just before each of these?
   a. △⬠△
   b. ☐◯
   c. △◯◯
   d. Penta-hen
   e. Bi-housand

Part C: Summary and Connections

When you return home, your professor asks you to write a paragraph or two summarizing your experiences with Hexaminian numeration. Specifically, she asks you to address the following questions:

1. How is the Hexaminian system like the decimal system? How is it different?

2. Did working in the Hexaminian system change your perspective on how numerals are written in the decimal system? If so, how?

Exploration 3.4 Base-Ten Blocks and the Decimal System

Purpose: Students use base-ten blocks to understand trading and the key features of the decimal system.

Materials: Base-ten blocks and the base-ten place value mat

Part A: Representing Numbers with Base-Ten Blocks

A common way to represent decimal numerals in the elementary grades is to place the appropriate base-ten blocks on a place value mat. For instance, 1,053 can be represented in the following way.

Thousands	Hundreds	Tens	Ones
(large cube)		(5 rods)	(3 units)

Use a minimal set of base-ten blocks to represent each numbers on a place value mat. When you are finished, draw a sketch of the base-ten blocks. Then complete the statement summarizing the different groups of ten used to form the number.

1. 296 = _____ thousands + _____ hundreds + _____ tens + _____ ones

2. 1,358 = _____ thousands + _____ hundreds + _____ tens + _____ ones

3. 2,206 = _____ thousands + _____ hundreds + _____ tens + _____ ones

Part B: Decomposing Numbers in Different Ways

In Part A, we decomposed each number using a minimal set of base-ten blocks. However, numbers can be decomposed in different ways by making exchanges between groups of ten.

1. Use a place value mat and base-ten blocks to represent each number in two ways that are different from the minimal set. When you are finished, draw a sketch of the base-ten blocks on the mat. Then complete the statements summarizing the different groups of ten used to form the number.

a. 296 = _____ thousands + _____ hundreds + _____ tens + _____ ones

296 = _____ thousands + _____ hundreds + _____ tens + _____ ones

b. 1,358 = _____ thousands + _____ hundreds + _____ tens + _____ ones

1,358 = _____ thousands + _____ hundreds + _____ tens + _____ ones

c. 2,206 = _____ thousands + _____ hundreds + _____ tens + _____ ones

2,206 = _____ thousands + _____ hundreds + _____ tens + _____ ones

2. a. Represent the number 3,061 in six different ways by filling in the blanks. Use exchanges rather than calculations to obtain your answers.

3,061 = _____ thousands + _____ hundreds + _____ tens + _____ ones

3,061 = _____ thousands + _____ hundreds + _____ tens + _____ ones

3,061 = _____ thousands + _____ hundreds + _____ tens + _____ ones

3,061 = _____ thousands + _____ hundreds + _____ tens + _____ ones

3,061 = _____ thousands + _____ hundreds + _____ tens + _____ ones

b. Why is it important to use the concept of exchanging rather than making calculations?

Part C: Composing Numbers

Numbers can also be composed by making exchanges to condense groups of ten. Make exchanges to rewrite each number using the fewest number of thousands, hundreds, tens, and ones.

1. 2 thousands + 12 hundreds + 35 tens + 24 ones

= _____ thousands + _____ hundreds + _____ tens + _____ ones

= _____

2. 5 thousands + 0 hundreds + 41 tens + 26 ones

= _____ thousands + _____ hundreds + _____ tens + _____ ones

= _____

3. 0 thousands + 56 hundreds + 0 tens + 534 ones

   = _____ thousands + _____ hundreds + _____ tens + _____ ones

   = _____

Part D: Summary and Connections

1. Many situations in our daily lives require us to make exchanges between groups of ten. List at least five examples of such situations.

2. In Part B, you decomposed 3,061 in different ways. Many elementary students struggle to decompose numbers in ways that are different from the minimal set. Why do you think this is so?

Exploration 3.5 Conceptualizing Large Numbers

Purpose: Students conceptualize the size of one million and one billion by visualizing these numbers in different ways.

Materials: A word processor and base-ten blocks

Part A: How Big is a Million?

Students often have trouble conceptualizing large numbers because they seldom have experience with these values. One easy way to conceptualize the relative size of large numbers is to use a word processor. For instance, consider the relative size of 1 to 1,000,000. Open a blank document on a word processor. On the first line of the page type 50 zeros. If you like, change the font or the size of the zeros so that they take up the entire first line. Copy and paste a line of zeros directly below the first one. You should now have 100 zeros arranged in the first two lines of the document.

1. a. If you copy the two lines, how many times will you need to paste the lines to generate 1,000 zeros?
   b. How many lines of print does it take to list 1,000 zeros? About how much of a page is this?

2. a. If you copy all 1,000 zeros, how many times will you need to paste the lines to generate 10,000 zeros?
   b. How many lines of print does it take to list 10,000 zeros? About how many pages is this?

3. a. If you copy the lines containing all 10,000 zeros, how many times will you need to paste them to generate 100,000 zeros?
   b. How many lines of print does it take to list 100,000 zeros? About how many pages is this?

4. a. Finally, if you copy the lines containing all 100,000 zeros, how many times will you need to paste them to generate 1,000,000 zeros?
   b. How many lines of print does it take to list 1,000,000 zeros? About how many pages is this?

5. Compare the amount of space required to type 1 zero to the amount of space required to type 1,000,000 zeros. What impressions does this make on you about the relative size of 1 to 1,000,000?

Part B: How Big is a Billion?

Another way to visualize the relative size of large numbers is to use base-ten blocks. For instance, we can make a group of 10,000 by making a **mega-long**, which is ten blocks stacked on top of one another. We can make a group of 100,000 by making a **mega-flat**, which is ten mega-longs placed side by side. If we set a unit next to these shapes, we can illustrate the relative size of 1 to 10,000 and 1 to 100,000.

1. With base-ten blocks, a group of 1,000,000 is represented with a **mega-block**.
    a. How would you use mega-flats to make a mega-block?
    b. In most cases, we would make a mega-block out of blocks, or groups of 1,000. How many blocks are needed to make a mega-block? This implies that 1,000,000 is equal to how many thousands?
    c. It is generally not cost effective to purchase enough blocks to make a mega-block. However, it is still possible to conceptualize the size of a mega-block by thinking about the dimensions of the unit cube. In most sets of base-ten blocks, the unit cube measures one centimeter on each side. If this is the case, what are the dimensions of one mega-block?
    d. Find something that has the approximate dimensions of a mega-block or build one out of cardboard. Set a single unit next to it. What impressions does this make on you about the relative size of 1 to 1,000,000?

2. a. How could you use base-ten blocks to represent a group of 10,000,000, a group of 100,000,000 and a group of 1,000,000,000?
    b. What shapes would each of the groups be? What dimensions would each shape have?
    c. Find something that has the approximate dimensions for each shape in your answer to the previous question. Set a single unit next to each object. What impressions does this make on you about the relative size of 1 to 10,000,000; 1 to 100,000,000; and 1 to 1,000,000,000?

Part C: Summary and Connections

1. a. Review your answers to Part A. What patterns do you see in the number of times you copied and pasted lines? What feature of the decimal system does this represent?
    b. What patterns do you see in the number of lines required to represent each number?

2. Review your answers to Part B. What patterns do you see in the shapes you used to represent the different groups of ten? How does this coincide with the periods of the decimal system?

Exploration 3.6 Understanding Base-*b* Numerals

Purpose: Students use bundled sticks to understand the key features of base-*b* numeration systems.

Materials: Popsicle sticks and rubber bands

Part A: Base five

In any base-*b* numeration system, the value of the base determines the size of the groups, the place values, and the digits used. Consequently, if we change the base from ten to another number greater than one, we get another numeration system that has all the features of the decimal system. For instance, base five is a grouping system in which we group by fives instead of tens. This means that the place values of base five include ones, fives, twenty-fives, and so on.

1. a. How many groups of five are there in one group of twenty-five?
   b. How many groups of twenty-five are there in one group of one hundred twenty-five?
   c. How many groups of five are there in one group of one hundred twenty-five?

2. Because we always regroup once we reach five groups of any one size, it is only necessary to use a few digits to represent whole numbers in base five. What are those digits going to be?

3. Make a set of bundled sticks that includes several groups for each place value in base five. Using the fewest bundles possible, represent each number with bundled sticks. Draw a sketch of the bundles you used, then write a statement summarizing the groups of five needed to form the number. The subscript indentifies the numeral as base five.
   a. 32_{five} b. 314_{five} c. 1320_{five}

Part B: Base Eight

In base eight, we group by eights rather than tens. This means the place values in base eight are ones, eights, sixty-fours and so on.

1. a. How many groups of eight are there in one group of sixty-four?
   b. How many groups of sixty-four are there in one group of five hundred twelve?
   c. How many groups of eight are there in one group of five hundred twelve?

2. Because we always regroup once we reach eight groups of any one size, it is only necessary to use a few digits to represent whole numbers in base eight. What are those digits going to be?

3. Make a set of bundled sticks that includes several groups for each place value in base eight. Using the fewest bundles possible, represent each number with bundled sticks. Draw a sketch of the bundles you used, then write a statement summarizing the groups of eight needed to form the number.

   a. 47_{eight} b. 207_{eight} c. 1355_{eight}

Part C: Summary and Connections

1. What role does zero play in base-b numeration?

2. How could you use bundled sticks to develop a procedure for converting numbers in base-b to base-ten? What about base-ten to base-b?

Chapter 4
Whole-Number Computation

Much of elementary mathematics revolves around the whole-number operations of addition, subtraction, multiplication, and division. These operations are important because they form the foundation for later work in mathematics and they are useful in our everyday lives. Children first come to understand the operations by manipulating and counting sets. Then, as they gain experience, they begin to see the operations in other ways and understand how they relate to one another. Children learn the properties of the operations and how to use them to make computations easier. They also learn how to apply the operations in a variety of contexts by solving story problems.

When learning to compute, children begin with the single-digit facts for addition and multiplication, and their counterparts for subtraction and division. After students have experience with these basic facts, they use them to compute with large numbers in a variety of ways. Such methods include manipulatives, written algorithms, calculators, mental computation, and estimation.

The explorations in this chapter investigate the many aspects of whole-number computation. Some activities consider the meaning of the operations, their properties, and the contexts in which they are used. Others look at the procedures for computing with large numbers by using manipulatives or by examining the computational strategies and mistakes of children. Specifically,

- Explorations 4.1 - 4.2 investigate the meanings and properties of addition and subtraction.
- Explorations 4.3 - 4.7 consider adding and subtracting with large numbers.
- Explorations 4.8 - 4.10 investigate the meanings and properties of multiplication and division.
- Explorations 4.11 - 4.15 consider multiplying and dividing large numbers.
- Explorations 4.16 - 4.17 explore computation in other numeration systems.

Exploration 4.1 Properties of Whole-Number Addition

Purpose: Students explore the properties of whole-number addition by using counters and an addition facts table.

Materials: Counters, paper, and pencil

Part A: The Closure Property of Whole-Number Addition

The **closure property of whole-number addition** states that the sum of any two whole numbers is a unique whole number.

1. Restate the closure property in your own words and give two numerical examples.

2. The property states that the sum is unique. What does this mean?

3. Can you use the definition of whole-number addition to explain why the closure property is true for whole-number sums?

4. In general, a set is said to be **closed** with respect to addition if the sum of every two elements in the set is an element in the set. Determine whether each set is closed with respect to addition. Explain how you know.
   a. $A = \{5, 10, 15, 20...\}$ b. $B = \{1, 2, 3, 4\}$

Part B: The Commutative Property of Whole-Number Addition

The **commutative property of whole-number addition** states that if a and b are any whole numbers, then $a + b = b + a$.

1. Restate the commutative property in your own words and give two numerical examples.

2. a. Draw two pictures that illustrate how to use the set model and the number line model to represent and solve $3 + 5 = 5 + 3$.
   b. Which model is most effective for illustrating the commutative property? Why?

3. After children have used models to help them understand addition, they often turn to a fact table to help them master all one hundred, single-digit addition facts.
   a. Consider the following table. Is the commutative property readily apparent? Explain how or why not.
   b. How might knowing the commutative property be useful to children when learning basic addition facts?

+	0	1	2	3	4	5	6	7	8	9
0	0	1	2	3	4	5	6	7	8	9
1	1	2	3	4	5	6	7	8	9	10
2	2	3	4	5	6	7	8	9	10	11
3	3	4	5	6	7	8	9	10	11	12
4	4	5	6	7	8	9	10	11	12	13
5	5	6	7	8	9	10	11	12	13	14
6	6	7	8	9	10	11	12	13	14	15
7	7	8	9	10	11	12	13	14	15	16
8	8	9	10	11	12	13	14	15	16	17
9	9	10	11	12	13	14	15	16	17	18

4. Can you use the definition of whole-number addition to explain why the commutative property is true for all whole-number sums?

Part C: The Associative Property of Whole-Number Addition

The **associative property of whole-number addition** states that if a, b, and c are any whole numbers, then $a + (b + c) = (a + b) + c$.

1. a. Restate the associative property in your own words and give two numerical examples.
   b. Why is the associative property useful when adding more than two numbers at once?

2. a. Draw two pictures that illustrate how to use the set model and the number line model to represent and solve $3 + (4 + 1) = (3 + 4) + 1$.
   b. Which model is most effective for illustrating the associative property? Why?

3. a. Is the associative property readily apparent in the addition facts table? Explain how or why not.
   b. How might knowing the associative property be useful to children when learning basic addition facts?

4. Can you use the definition of whole-number addition to explain why the associative property is true for all whole-number sums?

Part D: The Identity Property of Whole-Number Addition

The **identity property of whole-number addition** states that for any whole number a, there exists the unique number 0, called the **additive identity**, such that $a + 0 = a = 0 + a$.

1. a. Restate the identity property in your own words and give two numerical examples.
   b. What does it mean for the additive identity to be unique?

2. a. Draw two pictures that illustrate how to use the set model and the number line model to solve $0 + 8 = 8$.
   b. Which model is most effective for illustrating the associative property? Why?

3. a. Is the identity property readily apparent in the addition facts table? Explain how or why not.

   b. How might knowing the identity property be useful to children when learning basic addition facts?

4. Can you use the definition of whole-number addition to explain why the identity property is true for all whole-number sums?

Part E: Summary and Connections

1. The four properties given in this exploration are often called "algebraic properties." Why is it reasonable to label them this way?

2. Do any of these properties hold true for whole-number subtraction? Use numerical examples to explain your thinking.

Exploration 4.2 Addition and Subtraction Story Problems

Purpose: Students examine different types of word problems for addition and subtraction.

Materials: Counters

Part A: Modeling and Classifying Different Types of Word Problems

In the elementary classroom, students often learn different approaches to addition and subtraction by working with story problems. Whereas there is only one approach to addition, that of combining two disjoint sets and counting, there are three different approaches to subtraction. In the **take-away approach**, a certain number of objects are removed from a set and then the remaining objects are counted. In the **comparison approach**, two sets of objects are compared and any remaining unpaired objects are counted. In the **missing-addend** approach, a difference is computed by finding how much more is needed to go from one value to another.

Suppose each of the following problems is given to a kindergarten student who knows how to count but does not know any addition or subtraction facts. Use counters or a number line to show how the student might model each problem. Be sure to draw a picture of the student's potential solution. Next, decide if the problem is an addition or a subtraction problem. For subtraction problems, further classify them as indicating the take-away, the comparison, or the missing-addend approach. Finally, write the addition or subtraction fact represented in the problem. Use a question mark to represent any unknown value.

1. Juan had 8 cookies. He ate 2 of them. How many does he have left?

2. Jane had 7 stickers. She put some of them on a gift for her friend. She has 3 stickers left. How many stickers did she put on her friend's gift?

3. Erin had some stickers. Today Mia gave her 3 more stickers. Erin now has 9 stickers altogether. How many stickers did Erin have yesterday?

4. Ron has 4 dollars. He wants to buy a notebook that costs 6 dollars. How many more dollars does he need?

5. Nicki had 2 balloons. Her mother gave her 7 more balloons. How many balloons does Nicki have now?

6. Sam has 8 toy trucks. Ted has 3 more trucks than Sam. How many trucks does Ted have?

7. Katy has 5 candies and Leon has 2 candies. How many more candies does Katy have?

8. Bob has 7 cookies. Two are chocolate and the rest are peanut butter. How many peanut butter cookies does he have?

Part B: Writing Addition and Subtraction Story Problems

Part of the role of an elementary teacher is to help students learn how the approaches to addition and subtraction are reflected in story problems. This can entail writing story problems that represent a particular approach.

1. Write a short story problem appropriate for the elementary classroom that uses
   a. Addition with the problem $8 + 5$.
   b. The take-away approach with the difference $15 - 9$.
   c. The comparison approach with the difference $12 - 4$.
   d. The missing-addend approach with the difference $11 - 2$.

2. How was the language you used different between the addition problem and the subtraction problem using the take-away approach?

3. How was the language you used different between the addition problem and the subtraction problem using the missing-addend approach?

4. How was the language you used different between the subtraction problems using the take-away approach and the comparison approach?

Part C: Summary and Connections

1. Why is it useful to teach the approaches to addition and subtraction through story problems?

2. There are other ways to classify and think about story problems. For instance, the Cognitively Guided Instruction research classifies addition and subtraction story problems into four types; join, separate, compare, part-part-whole. Search the Internet to find information on this scheme for classifying problems. Use what you learn to reclassify the eight problems in Part A.

Exploration 4.3 Addition with Base-Ten Blocks

Purpose: Students examine the processes involved in adding large numbers by exploring ways of using base-ten blocks to add.

Materials: Base-ten blocks and a decimal place value mat

Part A: Your Method for Adding with Base-Ten Blocks

1. Devise your own method for using base-ten blocks to add two large numbers. Use it to compute the following sums. In each case, draw a picture that illustrates your method.
    a. 378 + 261
    b. 635 + 386

2. a. The definition of whole-number addition indicates that that a sum is computed by taking the union of two disjoint sets. How does your method use the union of two sets?
   b. How does your method use counting?

3. Regrouping is another important part of adding large numbers. How does your method make use of regrouping?

4. a. Does your method lead naturally to a written algorithm for adding two large numbers? If so, describe the steps in the written algorithm.
   b. Use your written algorithm to compute 241 + 169.

Part B: Base-Ten Blocks and the Standard Algorithm for Addition

1. Use the standard algorithm to compute the following sums with a pencil and paper. Next, determine how to use base-ten blocks and a place value mat to model each step in the algorithm. Draw a picture that illustrates the process.
    a. 337 + 284
    b. 1,057 + 583

2. a. How does the standard algorithm use the union of two sets? How is this represented with base-ten blocks?
   b. How does the standard algorithm make use of counting?
   c. How does the standard algorithm make use of regrouping? How is this represented with base-ten blocks?

Part C: Summary and Connections

Compare your method of adding with base-ten blocks to the standard algorithm. Use the similarities and differences you see to answer each question.

1. How does your method differ from the standard algorithm in the way it uses
    a. The union of two sets?
    b. Counting?
    c. Regrouping?

2. a. Which method of adding with base-ten blocks do you prefer: your method or the method used with the standard algorithm? Why?
   b. Which method do you think would be the easiest for children in the elementary grades to use and understand? Why?

3. a. Which written algorithm do you prefer: your method or the standard algorithm? Why?
   b. Which method do you think would be the easiest for children in the elementary grades to use and understand? Why?

Exploration 4.4 Subtraction with Base-Ten Blocks

Purpose: Students examine the processes involved in subtracting large numbers by exploring ways of using base-ten blocks to subtract.

Materials: Base-ten blocks and a decimal place value mat

Part A: Base-Ten Blocks and the Take-Away Approach

In the **take-away approach** to subtraction, a certain number of objects are removed from a set and the remaining objects are counted. This approach is often used when subtracting large numbers with base-ten blocks.

1. Devise your own method for using the take-away approach and base-ten blocks to subtract two large numbers. Use it to compute the following differences. In each case, draw a picture that illustrates your method.
   a. 547 – 285
   b. 1,091 – 674

2. How does your method make use of basic subtraction facts? How did you demonstrate these facts with base-ten blocks?

3. Exchanging one group from a higher place value for a group of ten in a lower place value is an important part of subtracting large numbers. How does your method make use of exchanges? How did you demonstrate this with the base-ten blocks?

4. Does your method of subtraction lead naturally to a written algorithm? If so, describe the steps in the algorithm and use it to compute 349 – 267. If not, explain why.

Part B: Base-Ten Blocks and the Comparison Approach

Another way to subtract is to use the **comparison approach**. In this approach, two sets are compared and any remaining unpaired objects are counted. Although less common, the comparison approach can also be used to subtract large numbers with base-ten blocks.

1. Devise your own method for using the comparison approach and base-ten blocks to subtract two large numbers. Use it to compute the following differences. In each case, draw a picture that illustrates your method.
   a. 437 – 268
   b. 1,123 – 862

2. How does your method make use of exchanges? How did you demonstrate this with the base-ten blocks?

3. Does your method of subtraction lead naturally to a written algorithm? If so, describe the steps in the algorithm and use it to compute 276 – 195. If not, explain why.

Part C: Subtracting with Base-Ten Blocks and the Standard Algorithm

1. Use the standard algorithm to compute the following differences with a pencil and paper. Next, determine how to use base-ten blocks and a place value mat to model each step in the algorithm. Draw a picture that illustrates the process.
   a. 479 – 238
   b. 526 – 387

2. How does the standard algorithm make use of basic subtraction facts? How did you demonstrate these facts with base-ten blocks?

3. How does the standard algorithm make use of exchanges? How did you demonstrate these facts with base-ten blocks?

Part D: Summary and Connections

Compare the three methods of subtracting large numbers. Use the similarities and differences you see to answer each question.

1. How are the three approaches similar or different in the way they use exchanges to compute a difference?

2. a. Which method of subtracting with base-ten blocks is most efficient? Why?
   b. Which method of subtracting with base-ten blocks do you prefer? Why?
   c. Which method would be the easiest for children in the elementary grades to use and understand? Why?

Exploration 4.5 Mental Addition and Subtraction

Purpose: Students practice adding and subtracting large numbers mentally.

Materials: Paper and pencil

Part A: Mental Addition

1. Find each sum mentally. If possible, avoid using the standard algorithm for addition. After computing each sum, describe the strategy you used.

   a.
   45
   +33

   b.
   156
   + 85

   c.
   268
   +139

   d.
   405
   +281

   e.
   2,567
   + 687

   f.
   4,156
   + 3,835

2. Examine the strategies you used. How did you make use of single-digit addition facts and properties of addition?

Part B: Mental Subtraction

1. Find each difference mentally. If possible, avoid using the standard algorithm for subtraction. After computing each difference, describe the strategy you used.

   a.
   53
   −21

   b.
   135
   − 48

   c.
   310
   −150

   d.
   506
   −381

   e.
   2,001
   − 798

   f.
   5,613
   −2,998

2. Examine the strategies you used. How did you make use of subtraction facts and properties of addition or subtraction?

Part C: Summary and Connections

1. Compare your strategies with one of your peers. Which of your strategies were the same? Which were different? Did there seem to be one strategy that you used most commonly?

2. How might examining your own strategies for mentally adding and subtracting affect how you teach mental addition and subtraction to your future students?

Exploration 4.6 Children's Strategies for Addition and Subtraction

Purpose: Students make sense of the strategies children use to add and subtract two-digit numbers.

Materials: Paper and pencil

Part A: Children's Strategies for Addition

Children often develop their own strategies for adding large numbers. As a teacher, you will need to decipher these strategies and determine whether they work for other sums.

1. To add 48 + 26, Michael first adds 48 + 20 = 68. He then adds 68 + 6 = 74.
    a. Describe Michael's strategy.
    b. Will it work for every sum? If so, use it to add 57 + 34.

2. Krupa used the following steps to add 48 + 26. First, she added 40 + 20 = 60. Next, she added 8 + 6 = 14. Finally, she added 60 + 14 = 74.
    a. Describe Krupa's strategy.
    b. Will it work for every sum? If so, use it to add 63 + 39.

3. Devin added 48 + 26 by subtracting 2 from 26 and adding it to 48. He then added 50 + 24 = 74.
    a. Describe Devin's strategy.
    b. Will it work for every sum? If so, use it to add 43 + 28.

4. Brianna adds 48 + 26 as follows: First, she adds 50 + 30 = 80, but then subtracts 2 to get 78 and 4 more to get 74.
    a. Describe Brianna's strategy.
    b. Will it work for every sum? If so, use it to add 37 + 27.

5. Reconsider the four strategies for addition. Do certain strategies work better with certain numbers? Explain.

Part B: Children's Strategies for Subtraction

Many children also develop their own strategies for subtracting large numbers.

1. To subtract 92 - 37, Sanjay first subtracts 92 − 30 = 62. He then subtracts 62 − 7 = 55.
    a. Describe Sanjay's strategy.
    b. Will it work for every difference? If so, use it to subtract 75 − 53.

2. Erika used the following steps to subtract 92 − 37. First, she subtracted 90 − 30 = 60. Next, she subtracted 60 − 7 = 53. Finally, she added 53 + 2 = 55.
    a. Describe Erika's strategy.
    b. Will it work for every difference? If so, use it to subtract 64 − 27.

3. Mei computed 92 − 37 by adding three to each number. She then subtracted 95 − 40 = 55.
    a. Describe Mei's strategy.
    b. Will it work for every difference? If so, use it to subtract 81 − 36.

4. Benjamin subtracted 92 − 37 in the following way. First, he subtracted 90 − 30 = 60. Next, he subtracted 2 − 7 = −5. Finally, he subtracted 60 − 5 = 55.
    a. Describe Benjamin's strategy.
    b. Will it work for every difference? If so, use it to subtract 62 − 18.

5. Reconsider the four strategies for subtraction. Do certain strategies work better with certain numbers? Explain

Part C: Summary and Connections

1. Reconsider the students' strategies for addition.
    a. Beyond regrouping and the use of single-digit addition facts, what other mathematical ideas did the students use in their strategies?
    b. Which of the strategies was most surprising? Why?

2. Reconsider the students' strategies for subtraction.
    a. Beyond exchanging and the use of basic subtraction facts, what other mathematical ideas do the students use in their strategies?
    b. Which of the strategies was most surprising? Why?

3. What are the advantages to letting children develop their own strategies for adding and subtracting large numbers? Are there are any disadvantages?

Exploration 4.7 Children's Errors with Addition and Subtraction

Purpose: Students examine common errors that children make when using the standard algorithms for addition and subtraction.

Materials: Paper and pencil

Part A: Children's Errors in the Standard Algorithm for Addition

Examine the set of problems solved by each student. In each case, identify the mistake, then explain the student's misunderstanding and what you might do to help the student.

1. Tamarcus:

    $$\begin{array}{r} 27 \\ +18 \\ \hline 35 \end{array} \qquad \begin{array}{r} 57 \\ +34 \\ \hline 81 \end{array} \qquad \begin{array}{r} 136 \\ +95 \\ \hline 121 \end{array}$$

2. Julianna:

    $$\begin{array}{r} \overset{1}{2}8 \\ +35 \\ \hline 531 \end{array} \qquad \begin{array}{r} \overset{1}{9}3 \\ +73 \\ \hline 67 \end{array} \qquad \begin{array}{r} 7\overset{1}{7} \\ +53 \\ \hline 211 \end{array}$$

3. Madeline:

    $$\begin{array}{r} 39 \\ +21 \\ \hline 510 \end{array} \qquad \begin{array}{r} 59 \\ +46 \\ \hline 915 \end{array} \qquad \begin{array}{r} 138 \\ +82 \\ \hline 11{,}110 \end{array}$$

4. Amado:

    $$\begin{array}{r} \overset{4}{2}8 \\ +16 \\ \hline 71 \end{array} \qquad \begin{array}{r} \overset{0}{5}7 \\ +23 \\ \hline 71 \end{array} \qquad \begin{array}{r} \overset{5}{1}64 \\ +91 \\ \hline 615 \end{array}$$

5. Akiko:

    $$\begin{array}{r} \overset{1}{2}9 \\ +14 \\ \hline 44 \end{array} \qquad \begin{array}{r} \overset{1}{3}7 \\ +29 \\ \hline 67 \end{array} \qquad \begin{array}{r} \overset{1}{1}29 \\ +53 \\ \hline 183 \end{array}$$

Part B: Children's Errors in the Standard Algorithm for Subtraction

Children can also have trouble with the standard algorithm for subtraction. Examine the problems solved by each student. In each case, identify the mistake, then explain the student's misunderstanding and what you might do to help the student.

1. Adalina:

$$\begin{array}{r} \overset{11}{3\cancel{1}} \\ -12 \\ \hline 29 \end{array} \qquad \begin{array}{r} \overset{16}{5\cancel{6}} \\ -38 \\ \hline 28 \end{array} \qquad \begin{array}{r} \overset{12}{4\cancel{2}} \\ -2\ 7 \\ \hline 2\ 5 \end{array}$$

2. Devon:

$$\begin{array}{r} \overset{3\ \ 13}{\cancel{4}\ \cancel{3}} \\ -2\ 9 \\ \hline 1\ 5 \end{array} \qquad \begin{array}{r} \overset{5\ \ 12}{\cancel{6}\ \cancel{2}} \\ -3\ 9 \\ \hline 2\ 4 \end{array} \qquad \begin{array}{r} \overset{0\ 10\ 16}{\cancel{1}\ \cancel{1}\ \cancel{6}} \\ -\ \ 7\ 9 \\ \hline 3\ 8 \end{array}$$

3. Jonathan:

$$\begin{array}{r} \overset{0\ 10\ 14}{\cancel{1}\ \cancel{0}\ \cancel{4}} \\ -\ \ 9\ 5 \\ \hline 1\ 9 \end{array} \qquad \begin{array}{r} \overset{2\ 10\ 16}{\cancel{3}\ \cancel{0}\ \cancel{6}} \\ -1\ 7\ 8 \\ \hline 1\ 3\ 8 \end{array} \qquad \begin{array}{r} \overset{4\ 10\ 11}{\cancel{5}\ \cancel{0}\ \cancel{1}} \\ -2\ 8\ 4 \\ \hline 2\ 2\ 7 \end{array}$$

4. Kelly:

$$\begin{array}{r} 23 \\ -17 \\ \hline 14 \end{array} \qquad \begin{array}{r} 75 \\ -28 \\ \hline 53 \end{array} \qquad \begin{array}{r} 142 \\ -\ 67 \\ \hline 125 \end{array}$$

5. LaShay:

$$\begin{array}{r} \overset{0\ \ 10}{\cancel{2}\ \cancel{0}\ \overset{12}{\cancel{2}}} \\ -\ \ 4\ 7 \\ \hline 6\ 5 \end{array} \qquad \begin{array}{r} \overset{1\ \ 10}{\cancel{3}\ \cancel{0}\ \overset{18}{\cancel{8}}} \\ -1\ 1\ 9 \\ \hline 9\ 9 \end{array} \qquad \begin{array}{r} \overset{5\ \ 10}{\cancel{7}\ \cancel{0}\ \overset{16}{\cancel{6}}} \\ -3\ 9\ 7 \\ \hline 2\ 1\ 9 \end{array}$$

Part C: Summary and Connections

1. Why are the standard algorithms for addition and subtraction difficult for many students to master? As a teacher, how might you make them easier for students to learn?

2. In daily life, it is more common to add and subtract large numbers with a calculator, yet far more classroom time is spent teaching written algorithms. What else might students learn from a written algorithm besides a way to make a computation?

Exploration 4.8 Properties of Whole-Number Multiplication

Purpose: Students explore the properties of whole-number multiplication by using colored chips and a multiplication facts table.

Materials: Counters, paper, and pencil

Part A: The Commutative Property of Whole-Number Multiplication

The **commutative property of whole-number multiplication** states that if a and b are any whole numbers, then $a \cdot b = b \cdot a$.

1. Restate the commutative property in your own words and give two numerical examples.

2. a. Draw three different pictures that illustrate how to use the set model, the array model, and the number line model to solve $3 \cdot 4 = 4 \cdot 3$.
   b. Which model is most effective for illustrating the commutative property? Why?

3. Is the commutative property readily apparent in the following multiplication facts table? Explain how or why not.

×	0	1	2	3	4	5	6	7	8	9
0	0	0	0	0	0	0	0	0	0	0
1	0	1	2	3	4	5	6	7	8	9
2	0	2	4	6	8	10	12	14	16	18
3	0	3	6	9	12	15	18	21	24	27
4	0	4	8	12	16	20	24	28	32	36
5	0	5	10	15	20	25	30	35	40	45
6	0	6	12	18	24	30	36	42	48	54
7	0	7	14	21	28	35	42	49	56	63
8	0	8	16	24	32	40	48	56	64	72
9	0	9	18	27	36	45	54	63	72	81

4. Use repeated addition to show that $3 \cdot 5 = 5 \cdot 3$.

Part B: The Associative Property of Whole-Number Multiplication

The associative **property of whole-number multiplication** states that if a, b, and c are any whole numbers, then $a \cdot (b \cdot c) = (a \cdot b) \cdot c$.

1. a. Restate the associative property in your own words and give two numerical examples.
   b. Why is the associative property useful when multiplying more than two numbers?

2. a. Draw three different pictures that illustrate how to use the set model, the array model, and the number line model to solve $2 \cdot (3 \cdot 4) = (2 \cdot 3) \cdot 4$.
   b. Which model is most effective for illustrating the associative property? Why?

3. Is the associative property readily apparent in the multiplication facts table? Explain how or why not.

Part C: The Identity Property of Whole-Number Multiplication

The **identity property of whole-number multiplication** states that for any whole number a, there exists the unique number 1, called the **multiplicative identity**, such that $a \cdot 1 = a$ and $1 \cdot a = a$.

1. a. Restate the identity property in your own words and give two numerical examples
   b. What does it mean for the multiplicative identity to be unique?
   c. Why is the multiplicative identity not the same as the additive identity?

2. a. Draw three different pictures that illustrate how to use the set model, the array model, and the number line model to solve $6 \cdot 1 = 6 = 1 \cdot 6$.
   b. Which model is most effective for illustrating the identity property? Why?

3. Is the identity property readily apparent in the multiplication facts table? Explain how or why not.

4. Use repeated addition to show that $4 \cdot 1 = 4$ and $1 \cdot 4 = 4$.

Part D: The Zero Multiplication Property of Whole-Number Multiplication

The **zero multiplication property of whole-number multiplication** states that for any whole number a, $a \cdot 0 = 0 = 0 \cdot a$.

1. Restate the zero multiplication property in your own words and give two numerical examples.

2. Is it possible to represent the zero multiplication property using the set model, the array model, or the number line model? Explain how or why not.

3. Is the zero multiplication property readily apparent in the multiplication facts table? Explain how or why not.

4. Use repeated addition to show that $5 \cdot 0 = 0 = 0 \cdot 5$.

Part E: The Distributive Property of Multiplication Over Addition

The **distributive property of multiplication over addition** states that for any whole numbers a, b, and c, $a(b + c) = a \cdot b + a \cdot c$.

1. a. Restate the distributive property of multiplication over addition in your own words and give two numerical examples.
   b. What is unique about the distributive property of multiplication over addition when compared to the other properties of multiplication?

2. a. Draw three different pictures that illustrate how to use the set model, the array model, and the number line model to solve $2(4+3) = 2 \cdot 4 + 2 \cdot 3$.
   b. Which model is most effective for illustrating the distributive property? Why?

3. a. Is there a distributive property of multiplication over subtraction? If so, write a statement of what you think the property would say.
   b. What concerns might there be with this property when using it with the whole numbers?

Part F: Summary and Connections

1. How might the properties of whole-number multiplication be useful to children when learning basic multiplication facts?

2. Do any of these properties hold true for whole-number division? Use numerical examples to explain your thinking.

Exploration 4.9 Multiplication and Division Story Problems

Purpose: Students examine different types of word problems for multiplication and division.

Materials: Counters, paper, pencil

Part A: Modeling and Classifying Different Types of Word Problems

In the elementary classroom, students often learn about the different approaches to multiplication and division by working with story problems. Two approaches are commonly used with multiplication. In the **repeated-addition approach**, a given number of objects or measures is combined to itself a given number of times. In the **Cartesian product approach**, each object in one set is paired with each object in another. There are also two approaches commonly used with division. In the **repeated-subtraction approach**, we find how many sets of a given size can be removed from a certain number of objects. In the **partitioning,** or **fair-shares approach**, we find how many objects are in each set when a certain number of objects are equally distributed to those sets.

Suppose each of the following problems is given to a third-grade student who knows how to add and subtract but has not mastered basic multiplication and division. Use counters or a number line to show how the student might model each problem. Be sure to draw a picture of the student's potential solution. Next, decide if the problem is a multiplication or a division problem. For multiplication problems, further classify them as using the repeated-addition or the Cartesian product approach. For division problems, further classify them as using the repeated-subtraction or the partitioning approach. Finally, write the multiplication or division fact represented in the problem. Use a question mark to represent any unknown value.

1. One piece of candy costs 25 cents. How much will three pieces cost?

2. If Celia walked 12 miles at a rate of 3 mph, how long did it take her?

3. If Marcy walks at 4 mph for 2 hours, how far will she have walked?

4. Each jar holds 8 ounces of liquid. If there are 42 ounces of juice in a pitcher, how many jars are needed to hold all the juice?

5. The Backyard Diner offers 3 main courses and 5 side dishes. A customer can order the Diner Special by selecting one main dish and one side dish. How many different Diner Specials are possible?

6. Kelly bought 5 pies at $8.50 each. How much did she spend?

7. Sue has 4 video games. Bill has 3 times as many games as Sue. How many games does Bill have?

8. Four children are planning to share a bag of 47 pieces of gumballs equally. How many pieces will each child get?

9. A rug measures 4 feet by 6 feet. What is the area of the rug?

10. Mark saved $560 this year. Last year he saved only $70. The amount he saved this year is how many times greater than the amount saved last year?

Part B: Writing Multiplication and Division Story Problems

Part of the role of an elementary teacher is to help students learn how the approaches to multiplication and division are reflected in story problems. This can entail writing story problems that represent a particular approach.

1. Write a short story problem appropriate for the elementary classroom that uses
    a. The repeated-addition approach to multiplication with the product $3 \cdot 4$.
    b. The Cartesian product approach to multiplication with the product $5 \cdot 2$.
    c. The repeated-subtraction approach to division with the quotient $12 \div 4$.
    d. The partitioning approach to division with the quotient $9 \div 3$.

2. How was the language you used different between the multiplication problems using the repeated-addition approach and the Cartesian product approach?

3. How was the language you used different between the division problems using the repeated-subtraction approach and the partitioning approach?

4. Which type of story problem was easiest for you to write? Most difficult? Why?

Part C: Summary and Connections

1. Why is it valuable to teach the different approaches to multiplication and division using story problems?

2. Which type of problem will be most difficult for elementary students? Why?

Exploration 4.10 Interpreting Remainders in Division

Purpose: Students use the context of a story problem to interpret the remainder of a quotient.

Materials: Paper and pencil

Part A: Interpreting the Remainder from a Division Story Problem

When a quotient occurs in a context, any remainder from the division must be interpreted in the context of the problem. Examine each division story problem and decide what happens with the remainder. After you are finished, write a short paragraph that summarizes the ways remainders can be handled in problem situations and the factors used to decide what to do with the remainders.

1. The fourth grade class is going on a trip. There are 32 students in the class. The principal requires that one adult for every five children must go on the trip. How many adults are needed?

2. You have 70 pencils to put in gift boxes. If each box contains 6 pencils, how many gift boxes can you make?

3. There are 24 large brownies on a tray. If you and 4 of your friends want to share the brownies equally, how many will each person get?

4. Emam has a board that is 18 feet long. How many 4-foot shelves can he cut from the board?

5. You have 25 books to share fairly with 8 children. How many will each child receive?

6. Jill bought 20 pounds of pears to make preserves. Each batch of preserves uses 3 pounds of pears. If she uses the left over pears to make a cobbler, how many pounds can she use in the cobbler?

7. Erin bought one can of chicken broth priced at 3 cans for $1.69. If tax is excluded, how much was she charged?

Part B: Writing Division Story Problems that Have a Remainder

Like other division story problems, elementary teachers also have to write division story problems that lead to remainders.

1. Write a short story problem appropriate for the elementary classroom that uses the quotient 26 ÷ 3 and has an answer of

    a. 8   b. 9   c. $8\frac{2}{3}$

2. How did the language in your story problems have to change for students to know what to do with the remainder?

Part C: Summary and Connections

1. Think of five or six situations in which you use division in your daily life. In each case, does the division lead to a remainder? If so, what do you do with it?

2. Why do you think it is important for children to learn how to solve division problems with remainders within a context?

Exploration 4.11 Multiplication with Base-Ten Blocks

Purpose: Students examine the processes involved in multiplying large numbers by exploring ways of using base-ten blocks to compute products.

Materials: Base-ten blocks and a decimal place value mat

Part A: Multiplying with Base-Ten Blocks and the Repeated-Addition Approach

Children in the elementary grades often learn to multiply using the **repeated-addition approach**; that is, the product of $a \cdot b$ is the sum of b added to itself a times. This approach is useful not only with single-digit products, but also with products involving large numbers.

1. Use repeated addition and base-ten blocks to develop your own method for multiplying two numbers. Use your method to compute the following products. In each case, draw a picture that illustrates your method.
   a. $5 \cdot 231$ b. $3 \cdot 438$

2. Regrouping is an important part of multiplying large numbers. How does your method make use of regrouping?

3. Although possible, repeated addition is seldom used when both factors in the product are large. Why?

Part B: Multiplying with Base-Ten Blocks and the Array Model

Another way to compute products with base-ten blocks is to use an extension of the array model. The next several questions demonstrate how.

1. Before the array can be made, first consider products involving different combinations of base-ten blocks. Complete the following table by giving the numerical product and the resulting base-ten block for each combination. The first one is given for you

Combination of Blocks	Numerical Product	Resulting Base-Ten Block
unit × unit	$1 \cdot 1 = 1$	A unit
unit × long or long × unit		
unit × flat or flat × unit		
long × long		
long × flat or flat × long		

2. The next step is complete the array. We begin by representing both factors with the appropriate number of blocks. List one set of blocks down the left side of the array and the other across the top. We then use the table in Question 1 to fill in the array. For example, the array for 35 · 23 is shown.

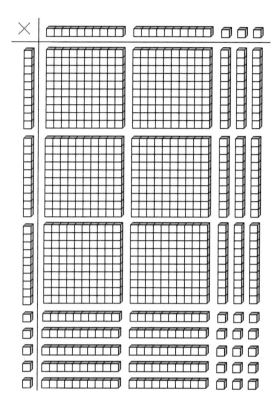

   a. Use a minimal set of base-ten blocks to make an array for each product. After you are finished, draw a picture of the array.
      i. 22 · 15   ii. 42 · 36
   b. Once an array is complete, what must be done to obtain the final product?

3. Does this method of multiplication lead naturally to a written algorithm? If so, describe the steps in the algorithm and use it to compute 87 · 43. If not, explain why.

Part C: Summary and Connections

Compare the methods of multiplying with base-ten blocks. Use the similarities and differences you see to answer each question.

1. How do the approaches differ in the way they use regrouping?

2. Which method do you think is the better way to multiply with base-ten blocks? Why?

3. What is the connection between the array model and the area of a rectangle?

Exploration 4.12 Division with Base-Ten Blocks

Purpose: Students examine the processes involved in dividing large numbers by exploring ways of using base-ten blocks to compute quotients.

Materials: Base-ten blocks

Part A: Base-Ten Blocks and the Repeated-Subtraction Approach

Base-ten blocks can be used in a number of ways to divide large numbers. One is to use the repeated-subtraction approach; that is, the quotient $a \div b$ is computed by finding the number of times b can be subtracted from a as long as $b \neq 0$.

1. Use repeated subtraction and base-ten blocks to develop your own method for dividing two numbers. Use your method to compute the following quotients. In each case, draw a picture that illustrates your method.
   a. $126 \div 14$ b. $296 \div 37$

2. Are exchanges necessary in your method? If so, how does your method use them?

3. Repeated subtraction is not always the most efficient way to divide. Describe the quotients for which it would be inefficient to use repeated subtraction?

Part B: Base-Ten Blocks and the Partitioning Approach

Base-ten blocks can also be used with the partitioning approach to division. In this approach, the quotient $a \div b$ is computed by finding the number of objects in each set when a objects are distributed equally among b sets.

1. Use the partitioning approach and base-ten blocks to develop your own method for dividing two numbers. Use your method to compute the following quotients. In each case, draw a picture that illustrates your method.
   a. $312 \div 8$ b. $564 \div 12$

2. Are exchanges necessary in your method? If so, how does your method use them?

3. The partitioning approach is not always efficient. Describe the quotients for which it would be inefficient to use the partitioning approach?

Part C: Base-Ten Blocks and the Array Model

Base-ten blocks and an extension of the array model are a third way to compute quotients. In this method, the dividend represents the total number of units in the array and the divisor represents one dimension of the array. The task then becomes to find the quotient, which is the other dimension of the array.

1. Consider the quotient of 495 ÷ 15. The following figure shows how to use flats, longs, and units to place 495 units into an array that has 15 units one dimension. What is the other dimension? In other words, what is the quotient when 495 is divided by 15?

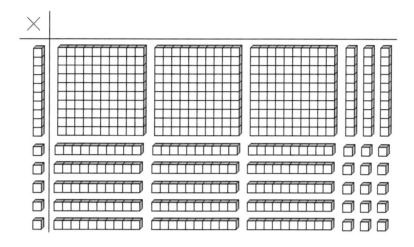

2. Use this method to compute the following quotients. Draw a picture of the array you used to make the computation.
   a. 209 ÷ 11   b. 624 ÷ 24

3. This method makes use of the missing-factor approach to division. What is this approach and how does the method make use of it?

4. Use an array to compute the quotient 281 ÷ 12, which has a remainder. How is the remainder illustrated in your model?

Part D: Summary and Connections

Compare the methods of dividing with base-ten blocks. Use the similarities and differences you see to answer each question.

1. How are the approaches similar in the way they make use of base-ten blocks? How are they different?

2. a. Which method do you think is the best way to divide with base-ten blocks? Why?
   b. Do you think it is reasonable to use this approach in the classroom? Why or why not?

3. Is it possible to represent the standard algorithm for division with base-ten blocks? If so, illustrate how.

Exploration 4.13 Mental Multiplication and Division

Purpose: Students practice multiplying and dividing large numbers mentally.

Materials: Paper and pencil

Part A: Mental Multiplication

1. Find each product mentally. If possible, avoid using the standard algorithm for multiplication. After computing each product, describe the strategy you used.

   a. 74 × 5
   b. 45 × 10
   c. 56 × 11
   d. 268 × 101
   e. 84 × 50
   f. 399 × 9

2. Examine the strategies you used. How did you make use of single-digit multiplication facts, properties of multiplication, and addition?

Part B: Mental Division

1. Find each quotient mentally. If possible avoid using the standard algorithm for division. After computing each quotient, describe the strategy you used.

   a. 2400 ÷ 30
   b. 1500 ÷ 60
   c. 2763 ÷ 9
   d. 1800 ÷ 45
   e. 399 ÷ 7

2. Examine the strategies you used. How did you use division facts and subtraction?

Part C: Summary and Connections

1. Compare your strategies with one of your peers. Which of your strategies were the same? Which were different? Did there seem to be one strategy that you used most commonly?

2. How might examining your own strategies for mentally multiplying and dividing affect how you teach mental multiplication and division to your future students?

Exploration 4.14 Children's Strategies for Multiplication and Division

Purpose: Students make sense of the strategies children use to multiply and divide multi-digit numbers.

Materials: Paper and pencil

Part A: Children's Strategies for Multiplication

As with addition and subtraction, children often develop their own strategies for multiplying large numbers. As a teacher, you will need to decipher these strategies and determine whether they work for other products.

1. Sara is given the following problem, "The bookstore received 35 boxes that contain 22 books each. How many books did they receive?" To solve the problem, she figures that 5 boxes must contain 110 books, so there must be 7 · 110, or 770 books.
   a. Describe Sara's strategy.
   b. Use her strategy if there were 45 boxes that contained 24 books each.
   c. Will her strategy work well with every product? Why or why not?

2. Devon is asked to solve the following problem: "There are 16 bags of candy with 28 pieces in each bag. How many pieces of candy are there in all?" He makes the following diagram to solve the problem. From it, he concludes that there are 448 pieces of candy.

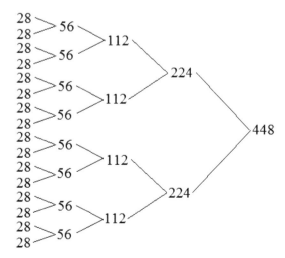

   a. Describe Devon's strategy.
   b. Use his strategy for 14 bags with 23 pieces of candy each.
   c. Will his strategy work well with every product? Why or why not?

3. Alissa is given the following problem: "There are 14 bags of candy with 26 pieces in each bag. How many pieces of candy are there in all?" She makes the following computations to solve the problem. She then claims that there are 364 pieces of candy.

$$10 \cdot 20 = 200$$
$$10 \cdot 6 = \underline{60}$$
$$260$$
$$4 \cdot 20 = 80$$
$$4 \cdot 6 = \underline{24}$$
$$104$$
$$\underline{260}$$
$$364$$

   a. Describe Alissa's strategy.
   b. Use her strategy for 23 bags of candy with 34 pieces each.
   c. Will her strategy always work? How do you know?

4. Marci is making necklaces to sell. Each necklace contains 58 beads, and she plans to make 32 necklaces. To find the number of beads she will need, she figures that 10 necklaces will need 580 beads, so 30 necklaces will need 3 · 580, or 1,740 beads. Marci then figures that 2 more necklaces will need 116 beads, so 32 necklaces will need 1,740 + 116, or 1,856 beads.
   a. Describe Marci's strategy.
   b. Use her strategy for 41 necklaces with 26 beads for each necklace.
   c. Will her strategy always work? How do you know?

Part B: Children's Strategies for Division

It is also common for children to develop their own strategies for dividing large numbers.

1. Ladarius is asked to solve the following problem: "I have 300 jelly beans. If I put 35 jelly beans in a small bag, how many bags can I make?" He makes the following computations to solve the problem. From them, he concludes that 8 bags can be filled with 20 jelly beans left over.

$$35 + 35 = 70$$
$$300 - 70 = 230$$
$$230 - 70 = 160$$
$$160 - 70 = 90$$
$$90 - 70 = 20$$

   a. Describe Ladarius' strategy.
   b. Use his strategy for 560 jelly beans with 60 jelly beans in each bag.
   c. Will his strategy always work? How do you know?

2. Pete is asked how many jelly beans each child will get if he equally shares 97 jelly beans with 4 children. He makes the following computations to solve the problem. From them, he claims that each child gets 24 jelly beans and 1 will be left over.

$$20 + 20 + 20 + 20 = 80$$
$$10 \div 4 = 2 \text{ R } 2$$
$$7 \div 4 = 1 \text{ R } 3$$
$$2 + 3 = 5$$
$$5 \div 4 = 1 \text{ R } 1$$

   a. Describe Pete's strategy.
   b. Use his strategy for 85 jelly beans shared equally with 3 children.
   c. Will his strategy always work? How do you know?

3. Samira is told that one jar holds 160 mL. She is asked how many jars can be filled with 2,100 mL of liquid. She makes the following computations to solve the problem. From them, she concludes that 13 jars each hold 160 mL of liquid and there is 20 mL of liquid left over.

$$10 \cdot 160 = 1,600$$
$$13 \cdot 160 = 2,080$$
$$14 \cdot 160 = 2,240$$

   a. Describe Samira's strategy.
   b. Use her strategy to find how many 150 mL jars can be filled with 2,400 mL of liquid.
   c. Will her strategy always work? How do you know?

Part C: Summary and Connections

1. Review the students' strategies for multiplication.
    a. Make a list of the mathematical ideas the students used in their strategies.
    b. Which strategy was most surprising to you? Why?

2. Review the students' strategies for division.
    a. Make a list of the mathematical ideas the students used in their strategies.
    b. Which strategy was most surprising to you? Why?

3. What are the advantages and disadvantages of allowing children develop their own strategies for multiplying and dividing larger numbers?

Exploration 4.15 Children's Errors with Multiplication and Division

Purpose: Students examine common errors that children make when using the standard algorithms for multiplication and division.

Materials: Paper and pencil

Part A: Children's Errors in the Standard Algorithm for Multiplication

Many children make mistakes when learning the standard algorithm for multiplication. Examine the set of problems solved by each student. In each case, identify the mistake, then explain the student's misunderstanding and what you might do to help the student.

1. James:

    $$\begin{array}{r} 34 \\ \times 15 \\ \hline 1700 \\ 3400 \\ \hline 5100 \end{array} \qquad \begin{array}{r} 45 \\ \times 22 \\ \hline 900 \\ 9000 \\ \hline 9900 \end{array} \qquad \begin{array}{r} 93 \\ \times 47 \\ \hline 6510 \\ 37200 \\ \hline 43710 \end{array}$$

2. Hunter:

    $$\begin{array}{r} 26 \\ \times 14 \\ \hline 824 \\ 260 \\ \hline 1084 \end{array} \qquad \begin{array}{r} 47 \\ \times 25 \\ \hline 2035 \\ 8140 \\ \hline 10175 \end{array} \qquad \begin{array}{r} 83 \\ \times 56 \\ \hline 4818 \\ 40150 \\ \hline 44968 \end{array}$$

3. Arianna:

    $$\begin{array}{r} 28 \\ \times 13 \\ \hline 84 \\ 28 \\ \hline 112 \end{array} \qquad \begin{array}{r} 47 \\ \times 33 \\ \hline 141 \\ 141 \\ \hline 282 \end{array} \qquad \begin{array}{r} 143 \\ \times 46 \\ \hline 858 \\ 572 \\ \hline 1430 \end{array}$$

4. Lucas:

    $$\begin{array}{r} 54 \\ \times 37 \\ \hline 162 \\ 3780 \\ \hline 3942 \end{array} \qquad \begin{array}{r} 67 \\ \times 29 \\ \hline 134 \\ 6030 \\ \hline 6164 \end{array} \qquad \begin{array}{r} 78 \\ \times 51 \\ \hline 390 \\ 780 \\ \hline 1172 \end{array}$$

Part B: Children's Errors in the Standard Algorithm for Division

Many children also have trouble with the standard algorithm for division. Examine the set of problems solved by each student. In each case, identify the mistake, then explain the student's misunderstanding and what you might do to help the student.

1. Lindsey:

$$
\begin{array}{r}
37 \\
5\overline{)1535} \\
\underline{15} \\
35 \\
\underline{35} \\
0
\end{array}
\qquad
\begin{array}{r}
31 \\
14\overline{)4214} \\
\underline{42} \\
14 \\
\underline{14} \\
0
\end{array}
\qquad
\begin{array}{r}
21 \\
17\overline{)3570} \\
\underline{34} \\
17 \\
\underline{17} \\
0
\end{array}
$$

2. Nadia:

$$
\begin{array}{r}
1317 \\
8\overline{)1176} \\
\underline{8} \\
37 \\
\underline{24} \\
136 \\
\underline{136} \\
0
\end{array}
\qquad
\begin{array}{r}
211 \\
17\overline{)527} \\
\underline{34} \\
187 \\
\underline{187} \\
0
\end{array}
\qquad
\begin{array}{r}
312 \\
23\overline{)966} \\
\underline{69} \\
276 \\
\underline{276} \\
0
\end{array}
$$

3. Thien:

$$
\begin{array}{r}
603 \\
6\overline{)1836} \\
\underline{18} \\
36 \\
\underline{36} \\
0
\end{array}
\qquad
\begin{array}{r}
612 \\
12\overline{)2592} \\
\underline{24} \\
19 \\
\underline{12} \\
72 \\
\underline{72} \\
0
\end{array}
\qquad
\begin{array}{r}
143 \\
16\overline{)5456} \\
\underline{48} \\
65 \\
\underline{64} \\
16 \\
\underline{16} \\
0
\end{array}
$$

Part C: Summary and Connections

1. Why are the standard algorithms for multiplication and division difficult for many students to master? As a teacher, what steps could you take to help students learn these algorithms more easily?

2. In daily life, it is more common to multiply and divide large numbers with a calculator, yet far more classroom time is spent teaching written algorithms. What else might students learn from a written algorithm besides how to make a computation?

Exploration 4.16 Hexaminian Addition and Subtraction

Purpose: Students learn to add and subtract in a new numeration system to reinforce concepts associated with whole-number addition and subtraction.

Materials: Counters, paper, and pencil

Part A: Adding in Hexaminian

On your last trip to Hexaminia (See Exploration 3.3), you and one of your classmates learned about Hexaminian numeration. Because you enjoyed yourself so much, you have decided to return to learn how to perform Hexaminian operations in the system. On the island, your Hexminian instructor greets you and agrees to teach you how to add in Hexaminian. Since your last visit, your teacher has learned much about the decimal system. He tells you that adding in Hexaminian is similar to adding in the decimal system.

1. Your instructor states that you must first learn to compute addition facts. This is done by counting out two sets with the appropriate number of objects, combining them together, and counting the total. For instance, the following diagram shows how to add △ + □.

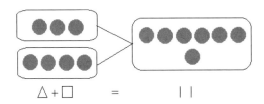

You are told to use counters and what you see in the diagram to compute the following sums. In each case, draw a picture of how you used the counters to compute the fact.
   a. △ + ○   b. V + ⬠   c. □ + | d. V + V

2. Your instructor tells you that the next step is to master all single-digit addition facts. To do so, he asks you to complete an addition fact table, which will be useful not only for computing sums with larger numbers but also for identifying properties of Hexaminian addition. Complete the following table and use it to answer each question. Use counters as needed.
   a. Is the Hexaminian system closed under addition? How do you know?
   b. Is Hexaminian addition commutative? How do you know?
   c. Is Hexaminian addition associative? If so, give an example. If not, give a counterexample.

d. Does Hexaminian addition have an additive identity? If so, what is it and how do you know?

+	○			V	△	□	⬠
○						
V						
△						
□						
⬠						

3. Your instructor sees you are progressing nicely and suggests that you try adding some larger numbers. He states that the Hexaminian algorithm for adding large numbers is similar to the standard algorithm for adding large decimal numbers. He tells you to use the table of addition facts to compute the following sums.

   a. □ | + | V   b. VV + | ⌂   c. △□ + V⌂   d. | V○ + □ △

Part B: Subtracting in Hexaminian

Because you have done well with addition, your instructor asks if you would like to try subtraction. He states that, like addition, subtracting in Hexaminian is similar to subtracting decimal numbers.

1. Your instructor demonstrates one way to think about subtraction by representing ⌂ − V with the following diagram:

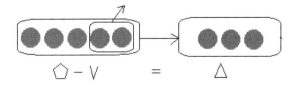

⌂ − V = △

a. What approach to subtraction did your instructor demonstrate?
b. Use counters and what you see in the diagram to compute the following subtraction facts. In each case, draw a picture of how you used the counters to compute the fact.

   i. ⌂ − △   ii. □ − | iii. ⌂ − ○   iv. △ − |

2. Your instructor says it is also possible to use addition to find the answer when subtracting. For instance, he says that $IV - \square = \square$ because $\square + \square = IV$.

   a. What approach to subtraction did your instructor use?

   b. Use this approach and the addition facts table completed earlier to compute the following subtraction facts.

   i. $II - \triangle$ ii. $I\triangle - \pentagon$ iii. $I\bigcirc - V$ iv. $I\bigcirc - \pentagon$

3. Your instructor sees you are progressing nicely and suggests that you try subtracting some larger numbers. Again, the Hexaminian algorithm for subtracting large numbers is similar to the standard algorithm for subtracting large decimal numbers. He tells you to use the table of addition facts to find the following differences.

   a. $\square\pentagon - V\square$

   b. $\triangle V - I\square$

   c. $I\square\bigcirc - \pentagon\pentagon$

   d. $I\bigcirc\bigcirc V - V\triangle\square$

Part C: Summary and Connections

When you return home, your professor asks you to write a paragraph or two summarizing your experiences with Hexaminian addition and subtraction. Specifically, she asks you to address the following two questions.

1. How are adding and subtracting in the Hexaminian system like adding and subtracting in the decimal system? How are they different?

2. Did adding and subtracting in the Hexaminian system help you gain a better understanding of these operations in the decimal system? If so, how?

Exploration 4.17 Hexaminian Multiplication and Division

Purpose: Students learn to multiply and divide in a new numeration system to reinforce concepts associated with whole-number multiplication and division.

Materials: Counters, paper, and pencil

Part A: Multiplying in Hexaminian

On your recent trip to Hexaminia, your instructor agrees to teach you not only addition and subtraction but also multiplication and division. He begins with Hexaminian multiplication and assures you it is very similar to multiplying in the decimal system.

1. Your instructor states that you must first learn to compute basic multiplication facts. He says you can interpret any product as adding the second factor to itself a number of times equal to the first factor. For instance, he represents △×⬠ using the following diagram:

   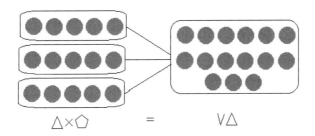

   You are told to use counters and what you see in the diagram to compute the following products. In each case, draw a picture of how you used the counters to compute the fact.

   a. △ × V   b. □ × I   c. □ × △   d. △ × ○

2. Your instructor tells you that the next step is to master all single-digit multiplication facts. To do so, he asks you to complete a multiplication fact table, which will be useful not only for computing products with larger numbers but also for identifying properties of Hexaminian multiplication. Complete the following table and use it to answer each question. Use counters as needed.
   a. Is the Hexaminian system closed under multiplication? How do you know?
   b. Is Hexaminian multiplication commutative? How do you know?
   c. Is Hexaminian multiplication associative? If so, give an example. If not, give a counterexample.
   d. Does the Hexaminian system have a multiplicative identity? If so, what is it and how do you know it is the multiplicative identity?

e. Does the Hexaminian system satisfy the zero multiplication property? If so, how do you know and what represents zero?

×	○			V	△	□	⬠
○						
V						
△						
□						
⬠						

3. Your teacher sees you are progressing nicely and suggests that you try multiplying some larger numbers. He states that the Hexaminian algorithm for multiplying large numbers is similar to the standard algorithm for multiplying large decimal numbers. He instructs you to use the table of multiplication facts to compute the following products.

   a. △△ × |V   b. V△ × |⬠   c. △□ × V○   d. V○⬠ × △|

Part B: Dividing in Hexaminian

Because you have done well with multiplication, your instructor asks if you would like to try division. He states that, like multiplication, dividing in Hexaminian is similar to dividing decimal numbers.

1. Your teacher demonstrates one way to think about division by representing |△ ÷ △ with the following diagram:

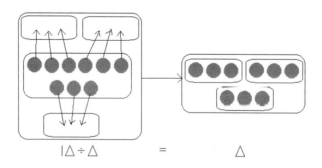

   a. What approach to division did your instructor demonstrate?
   b. Use counters and what you see in the diagram to compute the following division facts. In each case, draw a picture of how you used the counters to compute the fact.
      i. □ ÷ V   ii. ⬠ ÷ | iii. |□ ÷ ⬠   iv. ○ ÷ ⬠

2. Your instructor also says that it is possible to use multiplication to find the answer when dividing. For instance, he says that $\vee \triangle \div \bigcirc = \triangle$ because $\triangle \times \bigcirc = \vee \triangle$.
   a. What approach to division did your instructor use?
   b. Use this approach and the multiplication facts table completed earlier to compute the following division facts.
      i. $|\vee \div \square$ ii. $\vee \bigcirc \div \triangle$ iii. $|\square \div \vee$ iv. $\square | \div \bigcirc$

3. a. Is division by zero possible in the Hexaminian system? Explain.
   b. Does Hexaminian division ever have remainders? If so, give an example of the quotient of two numbers with a remainder.

4. Your instructor sees you are progressing nicely and suggests that you try dividing some larger numbers. Again, the Hexaminian algorithm for dividing large numbers is similar to the standard algorithm for dividing large decimal numbers. He instructs you to use the table of multiplication facts to find the following quotients.
   a. $|\vee \bigcirc \div \triangle$
   b. $\bigcirc | \div \square$
   c. $\bigcirc \triangle \vee \div | \square$
   d. $\vee \square \triangle \square \div \triangle \bigcirc$

Part C: Summary and Connections

When you return home, your professor asks you to write a paragraph or two summarizing your experiences with Hexaminian multiplication and division. Specifically, she asks you to address the following two questions.

1. How are multiplying and dividing in the Hexaminian system like multiplying and dividing in the decimal system? How are they different?

2. Did multiplying and dividing in the Hexaminian system help you gain a better understanding of these operations in the decimal system? If so, how?

Chapter 5
Basic Number Theory

Number theory is the study of numbers, their properties, and the relationships between them. Ideas from basic number theory play an important role in the elementary and middle grades. In the earliest grades, children use number theory to represent numbers in different ways and to classify numbers using simple characteristics, such as even and odd. In the late elementary and middle grades, children begin to learn about factors, multiples, and divisibility as they continue studying multiplication and division. These ideas are important not only because children use them to decompose numbers in different ways, but also because they are used in solving problems and in working with fractions and algebraic expressions.

Many of the explorations in Chapter 5 focus on divisibility and making sense of factors. Others address the greatest common factor and the least common multiple and how they can be used to solve problems. Specifically,

- Explorations 5.1 - 5.3 take an in-depth look at divisibility tests.
- Explorations 5.4 - 5.5 consider factors and prime numbers.
- Explorations 5.6 - 5.8 explore the greatest common factor and the least common multiple and how they are used to solve problems with cycles.

Exploration 5.1 Making Sense of Divisibility by 2, 5, and 10

Purpose: Students use counters and specific numbers to make sense of the divisibility tests for 2, 5, and 10.

Materials: Counters, paper, and pencil

Part A: Divisibility by 2, 5, and 10

The divisibility tests for 2, 5, and 10 are alike in that we only need to consider the ones digit to determine whether a number is divisible by any of the three numbers. To understand why the tests work, we must first understand what it means to be divisible by 2, 5, and 10.

1. The definition of divisibility states that if a and b are whole numbers, then a is **divisible** by b if and only if there exists a whole number c such that $a = b \cdot c$. Using this definition, what does it mean for a number to be divisible by
    a. 2?   b. 5?   c. 10?

2. How can you use counters to show that 10 is divisible by 2, 5 and 10? Draw diagrams to illustrate your method for showing divisibility by each number.

Part B: Multiples of 10 and Divisibility by 2, 5, and 10

In the last part, counters were used to show that 10 is divisible by 2, 5, and 10. Now consider multiples of 10.

1. a. What does it mean for a number to be a multiple of 10?
   b. Why are 20, 30, and 50 multiples of 10? Draw a diagram to justify your answer.
   c. If a number is a multiple of ten, why must it be divisible by 10?

2. How can you use counters and the fact that 20, 30, and 50 are multiples of 10 to determine whether these numbers are divisible by 2? Draw diagrams to illustrate your method.

3. Repeat question 2, but this time determine whether 20, 30, and 50 are divisible by 5.

4. Write a conjecture about multiples of 10 and divisibility by 2, 5, and 10. Test your conjecture with multiples of 10 larger than 100. If your conjecture is true for other multiples of 10, can you explain why?

Part C: Other Numbers and Divisibility by 2, 5, and 10

In Part B, counters were used to show that multiples of 10 are divisible by 2, 5, and 10. We now use this fact to check other numbers for divisibility by these values.

1. Before we check for divisibility by 2, 5, or 10, we must understand that any number can be written as a multiple of ten plus the digit in the ones place. For instance, 45 = 4 tens + 5 ones. Use this idea to fill in each blank with the appropriate number.

    a. 135 = _____ tens + _____ ones

    b. 478 = _____ tens + _____ ones

    c. 1,454 = _____ tens + _____ ones

    d. 3,258 = _____ tens + _____ ones

2. Next, consider divisibility by 2 and the fact that the sum of two numbers is divisible by 2 if each of the numbers is divisible by 2.
    a. What digits are divisible by 2?
    b. Any multiple of 10 is divisible by 2. Consequently, what single-digit numbers can be added to any multiple of 10 and still result in a number divisible by 2?
    c. What do your answers to the previous questions imply about divisibility by 2 and the ones digit of a number?

3. Now consider divisibility by 5 and the fact that the sum of two numbers is divisible by 5 if the numbers are divisible by 5.
    a. What digits are divisible by 5?
    b. Any multiple of 10 is divisible by 5. Consequently, what single-digit numbers can be added to any multiple of 10 and still result in a number divisible by 5?
    c. What do your answers to the previous questions imply about divisibility by 5 and the ones digit of a number?

4. Finally, consider divisibility by 10.
    a. What digits are divisible by 10?
    b. What do you notice about the ones digit of any multiple of 10?
    c. What do your answers to the previous questions imply about divisibility by 10 and the ones digit of a number?

Part D: Summary and Connections

1. Determine whether each number is divisible by 2, 5, or 10.
    a. 475
    b. 246
    c. 1,904
    d. 3,240

2. If possible, find a five-digit whole number with no repeated digits that is
    a. Divisible by 2 and 5, but not 10.
    b. Divisible by 2 and 10, but not 5.
    c. Divisible by 5 and 10, but not 2.
    d. Divisible by 2, 5, and 10.

3. a. If a number is divisible by 2 and 5, must it be divisible by 10? Why or why not?
   b. If a number is divisible by 10, must it be divisible by 2 and 5? Why or why not?

Exploration 5.2 Making Sense of Divisibility by 4 and 8

Purpose: Students use counters and specific numbers to make sense of the divisibility tests for 4 and 8.

Materials: Counters, paper, and pencil

Part A: Divisibility by 4

The divisibility test for 4 states that a number is divisible by 4 if, and only if, the last two digits represent a number divisible by 4. To understand why the test works, consider the following questions.

1. The definition of divisibility states that if a and b are whole numbers, then a is **divisible** by b if and only if there exists a whole number c such that $a = b \cdot c$. Using this definition, what does it mean for a number to be divisible by 4?

2. How can you use counters to determine whether 10 is divisible by 4? What about every multiple of 10? Draw diagrams to explain your answers.

3. How can you use counters to determine whether 100 is divisible by 4? What about every multiple of 100. Draw diagrams to explain your answers.

4. The last two questions illustrate that 4 does not divide every multiple of 10, but it does divide every multiple of 100. Consequently, to determine whether a number is divisible by 4, observe that any whole number can be written as the sum of a number of hundreds and 99 or fewer ones. For instance,

$$524 = 5 \text{ hundreds} + 24 \text{ ones}$$
$$1{,}518 = 15 \text{ hundreds} + 18 \text{ ones}$$

   a. Because 4 divides any multiple of 100, what is the only part of 524 that must be considered to determine whether it is divisible by 4?
   b. Likewise, what is the only part of 1,518 that must be considered to determine whether it is divisible by 4?
   c. What do your answers to the previous questions imply about divisibility by 4 and the last two digits of a number?

Part B: Divisibility by 8

The divisibility test for 8 states that a number is divisible by 8 if, and only if, the last three digits represent a number divisible by 8. To understand why the test works, consider the following questions.

1. Using the definition of divisibility, what does it mean for a number to be divisible by 8?

2. How can you use counters to determine whether 10 is divisible by 8? What about every multiple of 10? Draw diagrams to help you explain your answers.

3. How can you use counters to determine whether 100 is divisible by 8? What about every multiple of 100. Draw diagrams to explain your answers.

4. The last two questions illustrate that 8 does not divide every multiple of 10 or 100. However, 8 does divide every multiple of 1,000. How can you represent this?

5. In question 4 you showed that 8 divides every multiple of 1,000. Consequently, to determine whether a number is divisible by 8, we must observe that any whole number can be written as the sum of a number of thousands and 999 or fewer ones. For example,

    $$3{,}224 = 3 \text{ thousands} + 224 \text{ ones and}$$
    $$14{,}118 = 14 \text{ thousands} + 118 \text{ ones.}$$

    a. Because 8 divides any multiple of 1,000, what is the only part of 3,224 that must be considered to determine whether it is divisible by 8?
    b. Likewise, what is the only part of 14,118 that must be considered to determine whether it is divisible by 8?
    c. What do your answers to the previous questions imply about divisibility by 8 and the last three digits of a number?

Part C: Summary and Connections

1. Determine whether each number is divisible by 4 or 8.
    a. 1,332 b. 3,536 c. 7,514 d. 12,504

2. If possible, find a five-digit whole number with no repeated digits that is
    a. Divisible by 4, but not 8.
    b. Divisible by 8, but not 4.
    c. Divisible by 4 and 10.
    d. Divisible by 2, 4, and 5.

3. a. If a number is divisible by 4, must it be divisible by 8? If yes, explain why. If no, provide a counterexample.
   b. If a number is divisible by 8, must it be divisible by 4? If yes, explain why. If no, provide a counterexample.

Exploration 5.3 Making Sense of Divisibility by 3 and 9

Purpose: Students use counters and base-ten blocks to make sense of the divisibility tests for 3 and 9.

Materials: Counters, base-ten blocks, paper, and pencil

Part A: Divisibility by 3

The divisibility test for 3 states that a number is divisible by 3 if, and only if, the sum of its digits is divisible by 3. To understand why this test works, we must first understand which numbers are divisible by 3.

1. The definition of divisibility states that if a and b are whole numbers, then a is **divisible** by b if and only if there exists a whole number c such that $a = b \cdot c$. What does it mean for a number to be divisible by 3?

2. Determine whether each number is divisible by 3. In each case, draw a diagram that uses counters or base-ten blocks to explain your answer.
   a. 10   b. 100   c. 1,000   d. Any power of 10

3. The last question shows that no power of 10 is divisible by 3. However, every power of ten is close to a number that is. Find the number that is closest to each power of 10 and is divisible by 3. Draw a diagram using counters or base-ten blocks to show that this number is divisible by 3.
   a. 10   b. 100   c. 1,000   d. Any power of 10

4. Question 3 illustrates that every power of ten is one more than a number divisible by 3. For instance, 10 is one more than 9, 100 is one more than 99, and so on. This observation can be used to determine whether any number is divisible by 3. For instance, consider the following sequence of steps in which 1,326 is decomposed in a different way.

$$\begin{aligned} 1{,}326 &= 1 \cdot 1000 + 3 \cdot 100 + 2 \cdot 10 + 6 \\ &= 1 \cdot (999 + 1) + 3 \cdot (99 + 1) + 2 \cdot (9 + 1) + 6 \\ &= 1 \cdot 999 + 1 + 3 \cdot 99 + 3 + 2 \cdot 9 + 2 + 6 \\ &= [(1 \cdot 999) + (3 \cdot 99) + (2 \cdot 9)] + (1 + 3 + 2 + 6) \end{aligned}$$

   a. Provide a mathematical reason for each step taken in the decomposition.
   b. Because 999, 99, and 9 are divisible by 3, what must be true about the sum $[(1 \cdot 999) + (3 \cdot 99) + (2 \cdot 9)]$?
   c. Consider the sum $(1 + 3 + 2 + 6)$. What does the sum represent? Is it divisible by 3?
   d. If $[(1 \cdot 999) + (3 \cdot 99) + (2 \cdot 9)]$ is divisible by 3 and $(1 + 3 + 2 + 6)$ is divisible by 3, what can you conclude about the number 1,326?

5. What do the previous questions imply about divisibility by 3 and the digits of a number?

Part B: Divisibility by 9

The divisibility test for 9 is similar to the test for 3 in that a number is divisible by 9 if, and only if, the sum of its digits is divisible by 9.

1. Using the definition of divisibility, what does it mean for a number to be divisible by 9?

2. No power of 10 is divisible by 9, but every power 10 is close to a number that is. For each power of 10, find the number that is closest to it and is divisible by 9. Draw a diagram using counters or base-ten blocks to show that this number is divisible by 9.
   a. 10   b. 100   c. 1,000   d. Any power of 10

3. The observation that every power of 10 is one more than a number divisible by 9 can be used to determine whether any number is divisible by 9. For instance, consider the following sequence of steps in which 1,458 is decomposed.

$$\begin{aligned} 1{,}458 &= 1 \cdot 1000 + 4 \cdot 100 + 5 \cdot 10 + 8 \\ &= 1 \cdot (999 + 1) + 4 \cdot (99 + 1) + 5 \cdot (9 + 1) + 8 \\ &= 1 \cdot 999 + 1 + 4 \cdot 99 + 4 + 5 \cdot 9 + 5 + 8 \\ &= [(1 \cdot 999) + (4 \cdot 99) + (5 \cdot 9)] + (1 + 4 + 5 + 8) \end{aligned}$$

   a. Provide a mathematical reason for each step taken in the decomposition.
   b. Because 999, 99, and 9 are divisible by 9, what must be true about the sum $[(1 \cdot 999) + (4 \cdot 99) + (5 \cdot 9)]$?
   c. Consider the sum $(1 + 4 + 5 + 8)$. What does the sum represent? Is it divisible by 9?
   d. If $[(1 \cdot 999) + (3 \cdot 99) + (2 \cdot 9)]$ is divisible by 9 and $(1 + 3 + 2 + 6)$ is divisible by 9, what can you conclude about the number 1,326?

4. What do the previous questions imply about divisibility by 3 and the digits of a number?

Part C: Summary and Connections

1. Determine whether each number is divisible by 3 or 9.
   a. 159   b. 4,672   c. 8,565   d. 27,648

2. a. If a number is divisible by 3, must it be divisible by 9? If yes, explain why. If no, provide a counterexample.
   b. If a number is divisible by 9, must it be divisible by 3? If yes, explain why. If no, provide a counterexample.

Exploration 5.4 The Factor Game

Purpose: Students play a game to identify the properties of prime, composite, abundant, deficient, and perfect numbers.

Materials: Factor Game Board and colored pencils

Part A: Playing the Factor Game

Find a partner to play the factor game. The rules for the game are as follows:

- Player A selects a number from the Factor Game Board and circles it with a colored pencil. Player B then finds all factors of the number and circles them with a different color.
- Player B then selects and circles an unused number from the game board. Player A finds all available factors of the number and circles them. Once any number is circled, it cannot be used again.
- Players continue to take turns until there are no more numbers with available factors left on the board.
- **Illegal Move (Penalty Move):** If at any time a player chooses a number that has no available factors, that player has made an **illegal move**. Although the player making the illegal move gains the points, the player's next turn is forfeit. As a result, the other player selects the starting number for two turns in a row. If only illegal moves are possible, the game ends and no more points are awarded.
- At the end of the game, each player totals the numbers circled in his respective color. The player with the highest total is declared the winner.

Part B: Analyzing the Factor Game

Analyze the Factor Game by considering the following questions. In each case, explain your thinking.

1. Suppose you are the first person to have a turn in the Factor Game.
    a. Give at least 5 numbers that you can select that will allow your opponent to score only 1 point.
    b. Give at least 5 numbers that you can select that will allow your opponent to score more than 1 point.

2. Suppose you are the first person to have a turn in the Factor Game.
    a. What number can you select that will allow your opponent to score the same number of points as you? What other numbers accomplish the same purpose?
    b. What number can you select that will allow your opponent to score more points than you? What other numbers accomplish the same purpose?

3. In general, is it better to select a prime number or a composite number for a first move? Why?

4. Is electing to make an illegal move ever a winning strategy? If so, when?

Part C: Perfect, Deficient, and Abundant Numbers in the Factor Game

An interesting way to classify the natural numbers is to use the sums of their proper factors. A proper factor of a number is any factor that is not the number itself. For instance, the proper factors of 12 include 1, 2, 3, 4, and 6.

1. List the proper factors of each number.
    a. 6　　　b. 16　　　c. 24　　　d. 36　　　e. 45

2. Proper factors can be used to classify numbers as perfect, deficient, or abundant. A number is **perfect** if the sum of its proper factors is equal to the number. It is **deficient** if the sum of its proper factors is less than the number, and it is **abundant** if the sum of its proper factors is greater than the number. Determine whether each number is perfect, deficient, or abundant.
    a. 6　　　b. 16　　　c. 24　　　d. 36　　　e. 45

3. When playing the factor game, is it best to select a perfect number, a deficient number, or an abundant number as a first move? Which of these is the worst to select as a first move? Explain your answers.

4. Suppose you are the first player to take a turn in the factor game.
    a. What is the best number to select? Is it perfect, deficient, or abundant?
    b. What is the worst number to select? Is it perfect, deficient, or abundant?

Part D: Summary and Connections

1. If the Factor Game Board was changed to include the numbers from 1 to 75, would any of the strategies of the game change? Why or why not?

2. Do you think the Factor Game would be useful for teaching number theory in a fifth or sixth grade classroom? If so, describe how you would use it to do so.

FACTOR GAME BOARD

1	2	3	4	5
6	7	8	9	10
11	12	13	14	15
16	17	18	19	20
21	22	23	24	25
26	27	28	29	30
31	32	33	34	35
36	37	38	39	40
41	42	43	44	45
46	47	48	49	50

Exploration 5.5 Finding Primes to 200

Purpose: Students use the Sieve of Eratosthenes to find primes less than 200.

Materials: Paper and pencil

Part A: Finding Primes to 200

One way to find primes is to use the **Sieve of Eratosthenes**, a method that sifts out primes by looking at multiples of numbers rather than factors. To use the sieve, begin with a list of numbers and proceed in numerical order to cross out numbers that are not prime and circle those that are. For instance, consider the list of numbers up to 200 on the following page. First, cross out 1 because 1 is not prime. The next number, 2, is circled because it is prime. However, every multiple of 2 is composite, so cross out every even number remaining on the list. Next, return to the smallest number that has not been crossed out. Circle it as prime and then cross out all of its multiples. Continue to repeat the process until all the numbers in the list are either circled as prime or crossed out as composite. When you are finished, examine your results and answer the following questions.

1. Why is 2 the only even prime?

2. Why are 2 and 3 the only consecutive prime numbers?

3. What generalization can you make about the ones digit of every prime greater than 10?

4. Twin primes are primes that occur in pairs; that is, one prime is only two more than the other. How many pairs of twin primes are there between 1 and 200? List at least five of these pairs.

5. It is believed that every prime other than 2 and 3 may be one more or one less than a multiple of 6. Is this true for the primes less than 200?

6. The Sieve of Eratosthenes finds primes by eliminating multiples of numbers.
   a. What was the first number to fall through the sieve—that is, get crossed off—when you considered multiples of 2? What about multiples of 3? Multiples of 5?
   b. Is there a pattern to the first numbers that fall through the sieve? If so, what is it?
   c. When eliminating the multiples of a new prime, you may have noticed that there were numbers that were not crossed off or circled, yet had values smaller than the first number to fall through the sieve. What must be true about these numbers? How do you know?
   d. What is the largest prime you need to find before you know that all other uncircled numbers less than 200 must be prime?

Part B: Summary and Connections

1. If you were to find all the primes less than 500, what is the largest prime you would need to consider before you knew that all other uncircled numbers in the list are prime? What about primes less than 1,000? Primes less than n?

2. Write a short paragraph or two that describes how you might use an activity like this to teach a class of fifth graders about primes.

Numbers to 200

1	2	3	4	5	6	7	8	9	10
11	12	13	14	15	16	17	18	19	20
21	22	23	24	25	26	27	28	29	30
31	32	33	34	35	36	37	38	39	40
41	42	43	44	45	46	47	48	49	50
51	52	53	54	55	56	57	58	59	60
61	62	63	64	65	66	67	68	69	70
71	72	73	74	75	76	77	78	79	80
81	82	83	84	85	86	87	88	89	90
91	92	93	94	95	96	97	98	99	100
101	102	103	104	105	106	107	108	109	110
111	112	113	114	115	116	117	118	119	120
121	122	123	124	125	126	127	128	129	130
131	132	133	134	135	136	137	138	139	140
141	142	143	144	145	146	147	148	149	150
151	152	153	154	155	156	157	158	159	160
161	162	163	164	165	166	167	168	169	170
171	172	173	174	175	176	177	178	179	180
181	182	183	184	185	186	187	188	189	190
191	192	193	194	195	196	197	198	199	200

Exploration 5.6 Understanding the Prime Factorization Method for Finding the GCF and LCM

Purpose: Students use numbers to explore and understand why the prime factorization method can be used to find the GCF and LCM of two numbers.

Materials: Paper and pencil

Part A: Understanding the Prime Factorization Method for Finding the GCF

One way to find the GCF of two numbers is to use their prime factorizations. In this method, we find the prime factorization of each number and then compute the GCF by taking the product of every prime common to both factorizations. To make better sense of why this works, consider what happens when we use it to find the GCF(24, 36).

1. a. Find the prime factorization of 24.
   b. What primes occur in the prime factorization and how many times does each occur?

2. a. Find the prime factorization of 36.
   b. What primes occur in the prime factorization and how many times does each occur?

3. What primes do the prime factorizations of 24 and 36 have in common? Are there any primes that are common to the prime factorizations more than once?

4. The GCF of 24 and 36 is defined to be the largest factor common to both numbers.
   a. What are the factors of 24? Of 36?
   b. What factors do they have in common? What is the largest of these? This is the GCF of 24 and 36.

5. a. What is the prime factorization of the GCF of 24 and 36?
   b. How does the prime factorization of the GCF compare to the primes that 24 and 36 have in common?
   c. Using your observations from the previous questions, explain why it makes sense that the product of the primes that 24 and 36 have in common should be equal to the prime factorization of their GCF.

6. Use the prime factorization method to find the GCF of each pair of numbers.
   a. 180 and 432   b. 150 and 280   c. 520 and 546

Part B: Understanding the Prime Factorization Method for Finding the LCM

We can also use prime factorizations to find the LCM of two numbers. In this case, we find the prime factorization of each number and then compute the LCM by taking the product of the highest powers of any prime that occurs in either factorization. To make better sense of why this works, consider what happens when we use it to find the LCM(12, 18).

1. a. Find the prime factorization of 12.
    b. What primes occur in the prime factorization and how many times does each occur?
    c. Consider the multiples of 12. What primes must be common to all multiples of 12?

2. a. Find the prime factorization of 18.
    b. What primes occur in the prime factorization and how many times does each occur?
    c. Consider the multiples of 18. What primes must be common to all multiples of 18?

3. The LCM of 12 and 18 is defined to be the smallest multiple common to both. Use multiplication to list the first ten multiples of each number. What multiples do they have in common? What is the smallest of these? This is the LCM of 12 and 18.

4. a. What is the prime factorization of the LCM of 12 and 18?
    b. How does the prime factorization of the LCM compare to the highest powers of primes that occur in either of the factorizations of 12 and 18?
    c. Using your observations from the previous questions, explain why it makes sense that the product of the highest powers of the primes from the factorizations of either number should be equal to the prime factorization of their LCM.

5. Use the prime factorization method to find the LCM of each pair of numbers.
        a. 180 and 432   b. 150 and 280   c. 520 and 546

Part C: Summary and Connections

1. a. In question 6 of Part A, you found the GCF of three pairs of numbers. In question 5 of Part B, you found the LCM of the same three pairs of numbers. Do you see a connection between the GCF and the LCM of two numbers?
    b. Given two numbers, how are their GCF, their LCM, and their product related?

2. What is the GCF and the LCM of 24, 100, and 180? Explain how you can adapt the prime factorization method to find these numbers.

Exploration 5.7 Understanding the Euclidean Algorithm

Purpose: Students use numbers to explore and understand why the Euclidean algorithm can be used to find the GCF of two numbers.

Materials: Paper and pencil

Part A: Understanding the Euclidean Algorithm

The **Euclidean algorithm** uses long division and the division algorithm to find the GCF of two numbers. Specifically, it uses the following theorem:

Theorem: If a and b are any two nonzero whole numbers with $a \geq b$ and $a = bq + r$ for whole numbers q and r, with $r < b$, then $GCF(a, b) = GCF(b, r)$.

The theorem states that any common factors of a and b must also be common factors of b and r, where r is the remainder after a has been divided by b. Because this also includes the GCF, it must be that $GCF(a, b) = GCF(b, r)$. The Euclidean algorithm repeatedly uses this theorem to make a sequence of divisions in which the remainder of a quotient is divided into the previous divisor. In doing so, the numbers grow smaller until either we reach a GCF that is easy to find, or we get a remainder of 0. If we get a remainder of zero, we use the fact that the $GCF(n, 0) = n$. To make better sense of the Euclidean algorithm, consider what happens when we use it to find the $GCF(84, 60)$

1. Begin the algorithm by finding the remainder when the larger of two numbers is divided by the smaller. What is the remainder when 84 is divided by 60?

2. Next, consider the factors of 84 and 60.
    a. What are the factors of 84?
    b. What are the factors of 60?
    c. What factors do 84 and 60 have in common?

3. When we divide 84 by 60, we get a remainder of 24.
    a. What are the factors of 24?
    b. How do the factors of 24 compare to the common factors of 84 and 60?
    c. What does this imply about the $GCF(84, 60)$ and the $GCF(60, 24)$?

4. If necessary, repeat the process again to make the numbers even smaller. Specifically, divide 60 by 24 to get a remainder of 12.
    a. What are the factors of 60, 24, and 12?
    b. What factors do 60 and 24 have in common?
    c. How do the factors of 12 compare to the common factors of 60 and 24?
    d. What does this imply about the $GCF(60, 24)$ and the $GCF(24, 12)$? What about the $GCF(84, 60)$ and the $GCF(24, 12)$?

5. Finally, we complete the process.
   a. What is the GCF(24,12)?
   b. According to the Euclidean algorithm, what is the GCF(84,60)?

6. The Euclidean algorithm is based on the division algorithm, which states that if we divide two numbers, a and b, we get a quotient q and a remainder r, such that $a = bq + r$, where $0 \leq r < b$. If d is a factor of both a and b, write a short argument that explains why d is also a factor of the remainder r. (Hint: Rewrite the equation $a = bq + r$ in terms of r and then use properties of divisibility.)

Part B: Practicing the Euclidean Algorithm

Use the Euclidean algorithm to find the GCF of each pair of numbers.

1. GCF(120, 36)

2. GCF(805, 420)

3. GCF(8280, 8100)

Part C: Summary and Connections

1. What advantages does the Euclidean algorithm offer over other methods, such as the set intersection method or the prime factorization method, for finding the GCF of two numbers?

2. How can you use the Euclidean algorithm and the fact that $GCF(a, b) \cdot LCM(a, b) = a \cdot b$ to find the LCM of two large numbers?

3. How can a calculator with an integer divide key make the work involved in the Euclidean algorithm more efficient?

Exploration 5.8 Problems Involving Cycles

Purpose: Students use basic number theory to solve problems involving cycles.

Materials: Paper and pencil

Part A: Solving Problems that Involve Cycles

Cyclical situations are a common part of our everyday experience. They occur in our daily schedules, in our economy, in our biological rhythms, and even in the weather. The following problems involve cycles. As you solve them, think about how you are using ideas from basic number theory.

1. In North America, different species of cicadas can have different lengths of life cycles. One particular species hatches every 17 years and another hatches every 13. How often will both species hatch in the same year?

2. A certain bike has a total of seven gears. The two attached to the pedals have 44 and 56 teeth, respectively. The five attached to the rear tire have 14, 18, 22, 26, and 30 teeth, respectively. What is the fewest number of revolutions each gear has to make to return to their original position if the biker is using the gears with
    a. 44 and 30 teeth?　　　　　b. 44 and 22 teeth?
    c. 56 and 14 teeth?　　　　　d. 56 and 18 teeth?

3. Two consecutive stop lights on the main street of a town rotate through red, yellow, and green at different speeds. The first light turns green every 60 seconds and the second light turns green every 75 seconds. How many times during an hour will both lights turn green at the same time?

4. The time it takes for each of the four inner planets in our solar system to orbit the sun can be measured in earth days. Mercury orbits the sun in 88 earth days, Venus in 225, Earth in 365, and Mars in 687. If the four inner planets are aligned on a given day, how many earth days will pass before they are all aligned again? How long is this in years?

5. Many places along the Atlantic coast experience a high tide about every 12 hours and 25 minutes. On a given day in Savannah, GA the high tide occurs exactly at noon. How much time will pass before there is another high tide exactly at noon?

Part B: Summary and Connections

1. What common strategy did you use to solve the problems in this exploration?

2. Think of three other situations that involve cycles. Write a story problem for each situation that is appropriate for fifth or sixth grade students.

Chapter 6
The Integers

The previous three chapters focused on whole-number numeration and computation, two topics that encompass a large portion of elementary mathematics. Although an important set, the whole numbers do have several limitations. For instance, not every whole-number difference or quotient is defined. Likewise, whole numbers are inadequate for describing situations in which a connotation of direction is needed. We can remedy some of these limitations by expanding the set of whole numbers to the set of integers.

Students in the elementary grades are often introduced to the integers through familiar contexts like temperature and games. The primary goal is to give students a basic understanding of integers and their interpretations, so computation with integers is held to a minimum in these grades. Children learn more about integers in the middle grades, at which time they learn to order them, compute with them, and use them in problem-solving situations.

Many students find it difficult to work with integers because some of the properties and computations work in ways that are counterintuitive to what they learned with whole numbers. One way to help students get a better understanding of the integers and integer operations is to use concrete models. These models can be used to connect the integer operations to the whole-number operations and to show that the rules governing integer computations make sense.

Most of the explorations in Chapter 6 focus on understanding integers and integer operations by developing them through the use of concrete models. Specifically,

- Explorations 6.1 - 6.2 explore how to represent integers with the chip and number line models.
- Explorations 6.3 - 6.4 investigate how to use models to add and subtract integers.
- Explorations 6.5 - 6.6 examine integer multiplication and division.

Exploration 6.1 Integers and the Chip Model

Purpose: Students use variations of the chip model to represent integers.

Materials: Two-colored chips, paper, and colored pencils

Part A: The Chip Model

The chip model and its variations are a common way to represent integers. In the basic model, yellow chips represent positive integers and red chips represent negative integers. To show any particular integer, we put together the correct number of chips of the appropriate color. For instance, we represent

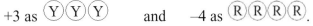

1. Use colored chips to represent each integer. In each case, use colored pencils to draw a diagram that illustrates the chips you used.
    a. +5   b. –3   c. +7   d. –5

2. a. How would you use colored chips to represent a pair of opposites such as +5 and –5?
   b. What would be the net effect of combining these chips together?

3. A combination of a positive chip and a negative chip is often called a "zero pair." Why does this name make sense?

Part B: The Charged Field Model

The charged field model is a variation of the chip model. In this case, integers are represented by using an appropriate number of positive (+) or negative charges (–).

1. Draw a diagram of how you would use the charged field model to represent each integer.
    a. –3   b. +4   c. –8   d. +6

2. What is the net value in each diagram?
    a.   b.   c.   d.
    + +   – – – –   + +   + + + + +
                                    – – – –   – – – –

3. How would you represent a zero pair in the charged field model?

Part C: The Hot and Cold Cubes Model

Another variation of the chip model uses hot and cold cubes. In this variation, think of a chef that is making soup with magical hot and cold cubes. If he adds a hot cube the temperature goes up one degree. If he adds a cold cube, the temperature goes down one degree.

1. The temperature of the soup is initially at zero. What integer would represent the temperature of the soup, if the chef puts in
   a. 4 cold cubes?   b. 8 cold cubes?   c. 5 hot cubes?   d. 10 hot cubes?

2. Give a verbal interpretation of each integer in terms of this model. How could you represent each integer with a picture?
   a. −3   b. +2   c. −6   d. +5

3. How would a zero pair be represented in the hot and cold cubes model?

Part D: The Golf Model

The game of golf offers a slightly different variation of the chip model. In this case, we use strokes above (+) or below (−) par to represent different integers.

1. Give the integer that represents each situation.
   a. 3 strokes over par   b. 2 strokes below par   c. par

2. Interpret each integer in terms of strokes above or below par.
   a. −2   b. +5   c. −4   d. +7

3. A golfer is at par. On the next two holes his caddie writes down −2 and +3. Where is the golfer now in relation to par?

4. Is it possible to represent a pair of opposites in the golf model? If so, how?

5. Are there any limitations to the integers that can be represented with this model? If so, what are they and why do they exist?

Part E: The Money Model

In the money model, integers are represented by deposits (+) or withdrawals (−) into a bank account.

1. Give the integer that represents each situation.
   a. Deposit of $25   b. Withdrawal of $10   c. Withdrawal of $165

2. Give a verbal interpretation of each integer in terms of the money model.
   a. −32   b. +17   c. −138   d. +245

3. If I open a bank account and make a deposit of $30 followed by a withdrawal of $10, what integer represents my current balance?

4. If (+20) represents my current balance, what combinations of deposits and withdrawals could lead to a balance of (–10)? Give at least 3 possible combinations.

5. Is it possible to represent a pair of opposites in the money model? If so, how?

Part F: Summary and Connections

1. Is it possible to represent the absolute value of an integer with the chip model or any of its variations? If so, how?

2. In what classroom situations do you think the chip or the charged field models would be the best way to represent the integers? What about the golf or money models?

3. Can you think of other situations that could be used as a variation of the chip model? If so, describe them.

Exploration 6.2 Integers and the Number Line Model

Purpose: Students use variations of the number line model to represent integers.

Materials: Paper and colored pencils

Part A: The Number Line Model

The number line model and its variations are a common way to represent integers. With a basic number line, each integer represents a distance and a direction relative to zero. Specifically, the value of the integer gives the distance from zero and the sign gives the direction, positive (+) to the right and negative (−) to the left.

1. State the position of each integer on a number with respect to zero.
   a. +4   b. −8   c. +6   d. −3

2. Zero is a special integer because it has no sign. What role does zero play in the set of integers and on the integer number line?

3. On a number line, two integers are **opposites** if they are an equal number of units from zero but are on opposite sides of zero. Identify the opposite of each integer and then draw a diagram that illustrates the integer and its opposite on a number line.
   a. −3   b. +5   c. −6   d. 0

4. The **absolute value** of an integer is the integer's distance from zero.
   a. Distance is always greater than or equal to zero. What does this imply about the absolute value of any integer?
   b. Find the absolute value of each integer and explain your thinking.
      i. −3   ii. +4   iii. −8   iv. +6
   c. What is the absolute value of zero? Explain your thinking.

Part B: The Thermometer Model

The thermometer model is a variation of the number line model.

1. a. How are positive integers represented on a thermometer? Negative integers?
   b. Why is the thermometer model a variation of the number line model?

2. What integer represents
   a. 60 degrees above zero?   b. 20 degrees below zero?   c. 35 degrees above zero?

3. Interpret each integer as a temperature above or below zero.
   a. −28   b. +41   c. −17   d. +157

3. Is it possible to represent two opposites with a thermometer? If so, how?

4. Does it matter whether a Fahrenheit thermometer or a Celsius thermometer is used? Explain.

Part C: The Elevator Model

Another variation of the number line model is the elevator model.

1. a. How are positive integers represented on an elevator? Negative integers?
   b. How is zero represented in this model?
   c. Why is the elevator model a variation of the number line model?

2. What integer represents
   a. 3 floors below ground level?   b. 6 floors above ground level?

3. Interpret each integer using the elevator model.
   a. –5   b. +3   c. –7   d. +37

4. Is it possible to represent two opposites with the elevator model? If so, how?

Part D: The Football Model

A third variation of the number line model is to use the results of plays as a football team moves up and down on a football field.

1. a. How are positive integers represented with football plays? Negative integers?
   b. Is zero represented in this model? If so, how?
   c. Why is the football model a variation of the number line model?

2. What integer represents a
   a. Gain of 12 yards?   b. Loss of 15 yards?   c. Loss of 10 yards?

3. Interpret each integer using the football model.
   a. +4   b. –6   c. +11   d. –12

4. Is it possible to represent two opposites with the football model? If so, how?

Part E: The Elevation Model

A final variation of the number line model is elevation.

1. a. How are positive integers represented with elevation? Negative integers?
   b. Is zero represented in this model? If so, how?
   c. Why is the elevation model a variation of the number line model?

2. What integer represents an elevation of
    a. 450 feet above sea level?   b. 96 feet below sea level?

3. Interpret each integer using the elevation model.
    a. +357   b. –217   c. +469   d. –133

4. Is it possible to represent two opposites with the elevation model? If so, how?

5. Does it matter whether the elevation is measured in feet or yards? Explain.

Part F: Summary and Connections

1. How can the absolute value of any integer be interpreted in each variation of the number line model?

2. Are there any limitations to the integers that can be represented with the different variations of the number line model? If so, what are the limitations and why do they exist?

3. How is the number line model different from the chip model in the way it represents the integers?

4. Can you think of other situations that could be used as a variation of the number line model? If so, describe them.

Exploration 6.3 Integer Addition and Subtraction with the Chip Model

Purpose: Students use the chip model to add and subtract integers.

Materials: Two-colored chips and colored pencils

Part A: Adding Integers with the Chip Model

One way to compute integer sums is to use the chip model and the intuitive idea behind whole-number addition. First represent each addend with the appropriate chips and then combine the chips together. However, before counting the total, make as many positive-negative pairs as possible. Each such pair has a net value of zero, so they are called **zero pairs** and they can be removed from consideration. The sum is then represented by the number and color of the remaining chips. For example, if yellow chips represent positive integers and red chips represent negative integers, then the sum $(-4) + 6 = 2$ is computed as shown.

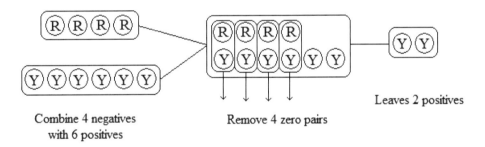

Combine 4 negatives with 6 positives   Remove 4 zero pairs   Leaves 2 positives

1. Use colored chips to compute each sum. In each case, use colored pencils to draw a diagram that illustrates how you solved the problem.
   a. $3 + 4$ b. $3 + (-5)$ c. $(-2) + 5$ d. $(-3) + (-5)$

2. Visualize how you would represent each sum with colored chips. Explain how you can use what you "see" in your mental picture to help you find the sum.
   a. $-150 + 200$ b. $-130 + (-450)$ c. $400 + (-625)$

3. a. What patterns do you see in the sums when the signs of the integers are the same?
   b. What patterns do you see in the sums when the signs of the integers are different?
   c. Using the patterns you found, write a definition for integer addition. Your definition should describe how to add two integers for any combination of signs.

3. a. Use the chip model to compute each sum. In each case, draw a diagram that illustrates how you solved the problem.
   i. $3 + (-3)$ ii. $(-4) + 4$ iii. $(-2) + 2$ iv. $5 + (5)$
   b. What property of integer addition does each of the sums illustrate?
   c. Use a number line to explain why this property of integer addition makes sense.

Part B: Subtracting Integers with the Chip Model

The chip model can also be used to subtract two integers. To do so, we use the intuitive idea behind the take-away approach to subtraction, but we must now take the signs of the integers into account. To begin, represent the first integer in the difference with the appropriate number and color of chip. Then, remove the appropriate number and color of chip representing the second integer. In many cases, there will not be enough of the right color of chip to remove. To remedy the situation, put in zero pairs until the correct number of chips is achieved. Once the appropriate chips have been removed, the difference is represented by the number and color of the remaining chips. For instance, the difference $3 - (-1) = 4$ is shown in the following diagram.

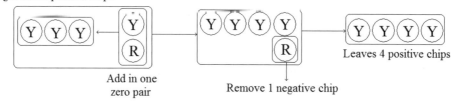

1. Use colored chips to compute each difference. In each case, use colored pencils to draw a diagram that illustrates how you solved the problem.
   a. $3 - (-2)$ b. $(-4) - 6$ c. $4 - 5$ d. $(-2) - (-5)$

2. Visualize how you would represent each difference with colored chips. Explain how you can use what you "see" in your mental picture to help you find the difference.
   a. $(-150) - (-100)$ b. $(-50) - 150$ c. $400 - 600$ d. $(-150) - (-200)$

3. It is not necessary to put in zero pairs in every difference. In which situations is it necessary and in which is it not?

4. a. What patterns do you see in the differences in which a positive is subtracted?
   b. What patterns do you see in the differences in which a negative is subtracted?
   c. Using the patterns you found, write a summary of how to subtract two integers.

5. When some students use colored chips to subtract integers, they first represent each integer in the difference with the appropriate chips. Then, they the chips for the second integer to reverse color, remove any zero pairs, and count the total.
   a. Use this method to compute the differences in question 1. Are your answers the same?
   b. What property of integer subtraction does this method use?

Part C: Summary and Connections

1. Why can it be helpful to show students how to add and subtract integers using a concrete model like the chip model before making them memorize rules for integer addition and subtraction?

2. When using a concrete model such as the chip model, why is it important to give problems such as the ones in question 2 of Part A and Part B?

Exploration 6.4 Integer Addition and Subtraction on a Number Line

Purpose: Students use a number line to add and subtract integers.

Materials: Paper and pencil

Part A: Adding Integers on a Number Line

Like whole-number addition, integer sums can be modeled and solved using a number line. Each integer in the sum is described as a person walking forward or backward a certain number of steps. The person begins at zero facing the positive direction. For a positive integer, the person walks forward that many units. For a negative integer, the person walks backward that many units. The sum is given by the final position of the person on the number line. For instance, to solve 5 + (–9), the following figure shows a person starting at zero facing the positive direction, moving forward 5 units, and then backward 9 units to arrive at (–4). Hence, 5 + (–9) = (–4).

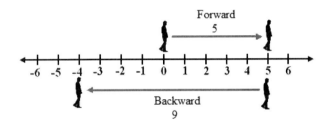

1. Use a number line to compute each sum. In each case, draw a diagram that illustrates how you solved the problem.
   a. 2 + 5   b. 4 + (–7)   c. (–6) + 8   d. (–5) + (–1)

2. a. What patterns do you see in the sums when the signs of the integers are the same?
   b. What patterns do you see in the sums when the signs of the integers are different?
   c. Using the patterns you found, write a definition for integer addition. Your definition should describe how to add two integers for any combination of signs.

3. a. Use a number line to compute each sum. In each case, draw a diagram that illustrates how you solved the problem.
      i. 2 + (–2)   ii. (–7) + 7   iii. (–6) + 6   iv. 5 + (–5)
   b. What property of integer addition does each of the sums illustrate?
   c. Use a number line to explain why this property of integer addition makes sense.

Part B: Subtracting Integers on a Number Line

Integer subtraction can also be modeled on a number line. As before, begin with a person at zero facing the positive direction. For a positive integer the person moves forward and for a negative integer the person moves backward. Now, however, the subtraction symbol tells the person to turn around and face the negative direction before making the second move. For instance, to solve 1 – (–5), the following figure shows a person starting at zero facing the positive direction and moving forward 1 unit. The person then turns around to face the negative direction and moves backward 5 units. In doing so, the person actually moves in the positive direction to arrive at 6. Hence, 1 – (–5) = 6.

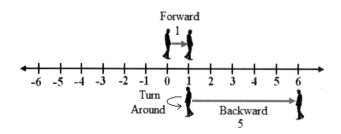

1. Use a number line to compute each problem. In each case, draw a diagram that illustrates how you solved the problem.
    a. 4 – 8   b. (–6) – 4   c. (–5) – (–4)   d. 6 – (–4)

2. a. What patterns do you see in the differences in which a positive is subtracted?
   b. What patterns do you see in the differences in which a negative is subtracted?
   c. Using the patterns you found, write a summary of how to subtract two integers.

Part C: Summary and Connections

1. One way to get students directly involved in using the number line model is to put a large number line on the floor along which they can move. Describe how a student would move along a number line to model and solve each of the following problems.
    a. 4 + 3   b. (–4) + 4   c. (–1) + (–8)   d. 5 + (–6)
    e. 2 – 8   f. (–2) – 5   g. (–3) – (–3)   h. 1 – (–7)

2. Design an activity appropriate for middle grades students in which they use the number line model to discover that the difference of two integers can be computed by adding the opposite of the second.

Exploration 6.5 Modeling Integer Multiplication

Purpose: Students use the chip and the number line models to multiply integers.

Materials: Two–colored chips, colored pencils, paper, and pencil

Part A: Multiplying Integers with the Chip Model

One way to represent integer multiplication is to use the chip model. To do so, we need to know how to interpret each factor and its sign. The first factor tells how many groups to put in (+) or take away (–). If chips need to be taken chips away, then it is necessary to put in zero pairs until the appropriate number of chips is reached. The second factor tells how many of what color of chip to put in each group. The product is represented by the remaining chips. For instance, to solve (–2) · (4), the following figure shows that we must first put in 2 · 4 = 8 zero pairs and then remove the necessary positive chips. Because 8 negative chips are left, (–2) · (4) = 8.

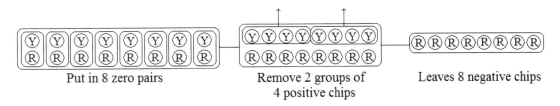

Put in 8 zero pairs   Remove 2 groups of 4 positive chips   Leaves 8 negative chips

1. Use the chip model to compute each product. In each case, draw a diagram that illustrates how you solved the problem.
   a. 6 · (–2)   b. (–3) · 1   c. (–2) · (–5)   d. 7 · 1

2. a. What patterns do you see in the products when the signs of the integers are the same?
   b. What patterns do you see in the products when the signs of the integers are different?
   c. Using the patterns you found, write a definition for integer multiplication. Your definition should describe how to multiply two integers for any combination of signs.

3. Do you think the chip model accurately represents the intuitive ideas of multiplication as a repeated sum or as the Cartesian product of two sets? Why or why not?

Part B: Multiplying Integers on a Number Line

Integer multiplication can also be computed on a number line, but to do so, we must again learn how to interpret the factors. As with addition and subtraction, imagine a person who starts at zero and moves back and forth along the number line. The first factor gives the direction in which the person is to face, positive (+) or negative (–), and the number of steps to be taken. The second factor gives the length of each step and whether the person moves forward (+) or backward (–). For instance, to solve (–3) · 4 the following figure shows that the person begins facing the negative direction and takes three steps forward, each step 4 units in length, to arrive at (–12). Hence, (–3) · 4 = (–12).

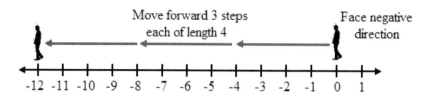

1. Use the number line model to compute each product. In each case, draw a diagram that illustrates how you solved the problem.

   a. 4 · (–2)   b. (–6) · 1   c. (–3) · (–5)   d. 2 · 7

2. a. What patterns do you see in the products when the signs of the two integers are the same?
   b. What patterns do you see in the products when the signs of the two integers are different?
   c. Are the patterns consistent with those you found in Part A?

3. Does the number line model accurately represents the intuitive ideas of multiplication as a repeated sum or as the Cartesian product of two sets? Why or why not?

Part C: Summary and Connections

1. Explain how you could adapt the number line model to use a car, its speed, and time traveled in the past or future to model integer multiplication. Specifically, describe the rules that the model would need to follow in order to compute integer products correctly.

2. Which model would be better to use with a group of children that was learning integer multiplication for the first time? Explain your preference.

Exploration 6.6 Exploring Integer Division

Purpose: Students use different approaches to understand integer division.

Materials: Two-colored chips, colored pencils, paper, and pencil

Part A: The Repeated-Subtraction Approach and Integer Division

In the **repeated-subtraction approach** to whole-number division, the quotient $a \div b$ is defined to be the number of times b can be subtracted from a as long as $b \neq 0$. The question might be raised whether this approach can also be applied to integer division. If a and b are positive integers, then $a \div b$ is a whole-number quotient. However, if a or b is negative, it is less clear whether the repeated-subtraction approach will work.

1. If possible, use colored chips and repeated subtraction to compute each quotient. When possible, draw a diagram that illustrates how you solved the problem.
   a. $9 \div 3$　　　b. $8 \div (-4)$　　　c. $(-12) \div 3$　　　d. $(-15) \div (-5)$

2. a. In the previous question, there was one quotient for each combination of signs. Does the repeated-subtraction approach work for every combination of signs? Why or why not?
   b. Why does repeated subtraction make sense in some cases but not in others?

3. For those quotients that could be computed with repeated subtraction, how did the signs of the factors compare to the sign of the quotient?

Part B: The Partitioning Approach and Integer Division

In the **partitioning approach** to whole-number division, the quotient $a \div b$ is the number of objects in each set after a objects have been equally distributed among b sets. Again, we check to see whether this approach can be applied to integer division.

1. If possible, use colored chips and partitioning to compute each quotient. When possible, draw a diagram that illustrates how you solved the problem.
   a. $6 \div 3$　　　b. $10 \div (-2)$　　　c. $(-14) \div 7$　　　d. $(-16) \div (-4)$

2. a. In the previous question, there was one quotient for each combination of signs. Does the partitioning approach work for every combination of signs? Why or why not?
   b. Why does the partitioning approach make sense in some cases but not in others?

3. For those quotients that could be computed through partitioning, how did the signs of the factors compare to the sign of the quotient?

Part C: The Missing-Factor Approach and Integer Division

In the **missing-factor approach** to whole-number division, the quotient $a \div b = c$ if, and only if, $a = b \cdot c$ for some whole number c. To use this approach, change the quotient into a product in which one of the factors is unknown. We then use our knowledge of multiplication to find the missing number. Again, we check to see whether this approach can be applied to integer division.

1. Write each quotient as a product in which one factor is missing. When possible, use your knowledge of integer multiplication to find the missing number.
   a. $21 \div 3$ b. $18 \div (-9)$ c. $(-25) \div 5$ d. $(-24) \div (-8)$

2. In the previous question, there was one quotient for each combination of signs. Does the missing-factor approach work for every combination of signs? Why or why not?

3. For those quotients that could be computed with the missing-factor approach, how did the signs of the factors compare to the sign of the quotient?

4. Can you use the missing-factor approach to find the quotient $(-16) \div 7$? Why or why not?

Part D: Summary and Connections

1. Of the three approaches, which was the only one to work for every combination of signs?

2. Use what you have learned to write several sentences that summarize how the signs of the factors compare to the sign of the quotient?

3. a. Even though not every approach worked for every combination of signs, do you think that one was more intuitive to use than the others? If so, why?
   b. How might your answer to the previous question influence how you might teach integer division?

Chapter 7
Fractions and the Rational Numbers

In Chapter 6, the set of whole numbers was extended to the integers. Like the whole numbers, the integers are not sufficient for handling every mathematical situation. For instance, not every integer quotient is defined. Likewise, the integers are inadequate for describing part-to-whole relationships. We can remedy these limitations by extending the integers to the set of rational numbers.

In the elementary grades, students first learn about the rational numbers through their work with common fractions, such as $\frac{1}{2}$ and $\frac{2}{3}$. Students learn to interpret the meaning of these fractions by using them to represent part-to-whole relationships, often with the use of concrete models. From the third grade on, children learn to represent, order, and compute with fractions. As they do, they learn that fractions not only represent part-to-whole relationships but also represent measures, ratios, and even quotients.

Developing a solid foundation with fractions is important because without it, fractions can become a serious obstacle in a child's mathematical development. As with the integers, many of the concepts and operations with fractions are counterintuitive to what students learn with the whole numbers. For instance, many children struggle with the fact that different fractions can have the same value and that they can make changes between these fractions once they decide which one best serves their needs. Children can also struggle to work correctly with the numerators and denominators of fractions when computing with them.

The explorations in Chapter 7 primarily focus on developing a conceptual understanding of the rational numbers and fraction operations. Specifically,

- Explorations 7.1 - 7.4 consider ways of representing fractions and student problems in understanding them.
- Explorations 7.5 - 7.8 look at rational number addition and subtraction and student misconceptions about these operations.
- Explorations 7. 9 - 7.14 investigate rational number multiplication and division and the problems students have in making sense of these operations.

Exploration 7.1 Exploring Fractions with the Area Model

Purpose: Students use versions of the area model to understand fractions.

Materials: Fraction disks, pattern blocks, and a 5 × 5 geoboard

Part A: Fraction Disks

In the **area**, or **regions model**, the whole is represented with the area of a shape that can be divided into equivalent sub-regions. It is a particularly powerful model because it allows us to count and see the part-to-whole relationship directly. Fraction disks are a common version of the area model that uses the area of a circle to represent the unit. Some examples of fractions represented with fraction disks are shown. Use a set of fraction disks to answer each question.

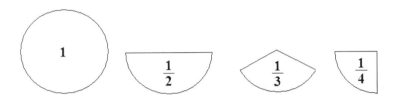

1. Suppose the complete circle represents 1. Represent each fraction with fraction disks and then sketch the fraction disks you used. What part of the fraction told you which pieces to use? What part told you how many pieces to use?

    a. $\frac{2}{3}$ b. $\frac{3}{8}$ c. $1\frac{1}{2}$ d. $2\frac{1}{3}$

2. If the half circle was the unit, what would be the fractional values of the other pieces?

3. Two fractions are **equivalent** if they have the same numerical value, but different part-to-whole interpretations.

    a. How can you use fraction disks to show that $\frac{2}{3}$ and $\frac{4}{6}$ are equivalent?

    b. A standard set of fraction disks contains pieces that represent one, halves, thirds, fourths, sixths, and eighths. Make a list of the equivalent fractions in a standard set of fraction disks.

4. a. Draw a diagram that uses fractions disks to show $\frac{5}{3}$ is equivalent to $1\frac{2}{3}$.

    b. Draw a diagram that uses fractions disks to show $2\frac{1}{4}$ is equivalent to $\frac{9}{4}$.

    c. How could you use your diagrams to show students how to convert improper fractions to mixed numbers and vice versa?

Part B: Pattern Blocks

Consider the following four shapes from a set of pattern blocks. Answer each question based on the fractional relationships between them.

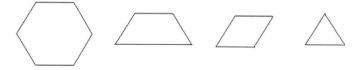

1. What shape is
    a. $\frac{1}{2}$ of the hexagon?
    b. $\frac{1}{3}$ of the hexagon?
    c. $\frac{1}{6}$ of the hexagon?
    d. $\frac{1}{3}$ of the trapezoid?
    e. $\frac{2}{3}$ of the trapezoid?
    f. $\frac{1}{2}$ of the rhombus?

2. a. Suppose ⬡⬡ represents 1. What fraction does each of the following represent?
    i. a trapezoid
    ii. a rhombus
    iii. a triangle
    iv. two rhombi
    v. three trapezoids
    vi. three hexagons
    vii. fifteen triangles
    viii. five rhombi

   b. Sketch three combinations of shapes that are equivalent to $\frac{3}{4}$.

3. a. Suppose a trapezoid represents $\frac{2}{3}$. Sketch a combination of shapes that represents 1.
   b. Suppose a hexagon represents $\frac{3}{4}$. Sketch a combination of shapes that represents 1.
   c. Suppose a rhombus represents $\frac{1}{2}$. Sketch a combination of shapes that represents 1.

4. Which is bigger, $\frac{3}{4}$ or $\frac{5}{6}$? Explain how you used pattern blocks to decide.

Part C: Geoboard

1. Suppose the largest square on a 5 × 5 geoboard represents 1. How many different ways can you use rubber bands to represent $\frac{1}{2}$ on the geoboard? Be creative.

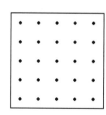

2. Suppose the largest square on a 5 × 5 geoboard represents 1. Use rubber bands to make a shape that has each fractional value.

   a. $\dfrac{1}{16}$ b. $\dfrac{1}{8}$ c. $\dfrac{1}{4}$ d. $\dfrac{1}{32}$ e. $\dfrac{5}{8}$ f. $\dfrac{15}{32}$

3. How could you use a geoboard(s) to show that

   a. $\dfrac{2}{4}$ is equivalent to $\dfrac{1}{2}$? b. $1\dfrac{3}{4}$ is equivalent to $\dfrac{7}{4}$? c. $\dfrac{3}{8}$ is less than $\dfrac{9}{16}$?

Part C: Summary and Connections

1. Describe how each manipulative uses area to represent fractions.
   a. Fraction disks   b. Pattern blocks   c. Geoboard

2. a. Why do you think it is important to use more than one area model to represent fractions?
   b. Use one manipulative from this exploration to design an activity that is appropriate for the elementary grades and in which children must use the manipulative to answers questions about fractions.

138   Chapter 7 Fractions and the Rational Numbers

Exploration 7.2 Exploring Fractions with the Length Model

Purpose: Students use different versions of the length model to understand fractions.

Materials: Fraction bars, colored rods, paper, and pencil

Part A: Fraction Bars

In the **length**, or **measurement** model, some length is used as the unit. With fraction bars, we use the length of a rectangle as one and then subdivide it into appropriate sublengths to represent particular fractions. A standard set of fraction bars has lengths that represent one, halves, thirds, fourths, fifths, sixths, eighths, tenths, and twelfths.

1. Let the length of the longest rectangle represent 1. Represent each fraction with fraction bars and then draw a diagram that illustrates the fraction bars you used. What part of the fraction told you which pieces to use? What part told you how many pieces to use?

    a. $\dfrac{3}{4}$ b. $\dfrac{5}{8}$ c. $\dfrac{6}{5}$ d. $1\dfrac{7}{12}$

2. Two fractions are **equivalent** if they have the same numerical value but different part-to-whole interpretations.

    a. How can you use fraction disks to show that $\dfrac{2}{3}$ and $\dfrac{4}{6}$ are equivalent?

    b. Make a list of the equivalent fractions in a standard set of fraction bars.

3. a. How can you use fraction bars to determine whether one fraction is less than another?
   b. Use fraction bars to determine which fraction in each pair is smaller. In each case, draw a diagram showing how you used fraction bars to make your determination.

    i. $\dfrac{2}{3}$ and $\dfrac{5}{6}$ ii. $\dfrac{7}{8}$ and $\dfrac{9}{12}$ iii. $\dfrac{3}{5}$ and $\dfrac{7}{10}$

Part B: Colored Rods

Colored rods are another manipulative that can be used to represent fractions with the length model. Use a standard set of colored rods containing ten different colors and lengths to answer each question.

1. If the brown rod is 1, what fractional part of the brown rod is each rod?
    a. White   b. Red   c. Purple   d. Dark green

2. If the blue rod is 1, what fractional part of the blue rod is each rod?
    a. Light green   b. Dark green   c. Purple   d. Orange

3. If the light green rod has a value of $\frac{1}{2}$, which rod has the given value?

   a. 1   b. $\frac{1}{3}$ c. $\frac{3}{2}$

4. If the blue rod has a value of $\frac{9}{10}$, which rod has the given value?

   a. 1   b. $\frac{1}{2}$ c. $\frac{1}{5}$

5. If the red rod has a value of $\frac{1}{4}$, which rod has the given value?

   a. 1   b. $\frac{1}{2}$ c. $\frac{3}{4}$

Part C: Number Line

When using a number line, the **unit** is always the distance from 0 to 1. Use a number line to answer each question.

1. Plot each point on the number line. Label them with the given capital letter.

   A. $\frac{2}{3}$ B. $\frac{3}{4}$ C. $\frac{11}{3}$ D. $\frac{-5}{2}$ E. $-1\frac{3}{4}$ F. $2\frac{2}{3}$

   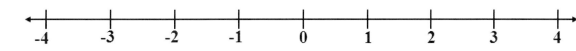

2. Mark the approximate location of 1 on each number line.

   a.

   b.
   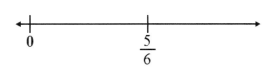

   c.
   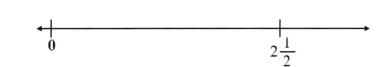

3. Consider the distance from A to B. Find the point located $\frac{1}{4}$ of the distance from A to B. What number is associated with that point?

   a.
   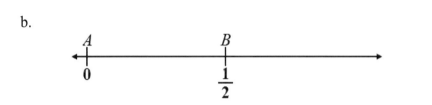

   b.

   c.

4. On which number line is segment $\overline{AB}$ the longest? How do you know?

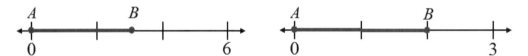

Part D: Summary and Connections

1. What does this exploration reveal about the difference between the unit and the whole?

2. a. Of the three variations to the length model, which was the most useful to you in representing fractions? Why?
   b. Use one manipulative from this exploration to design an activity that is appropriate for the elementary grades and in which children must use the manipulative to answers questions about fractions.

3. How could you use fraction bars to teach a group of children how to simplify fractions?

Exploration 7.3 Exploring Fractions with the Set Model

Purpose: Students use the set model to understand fractions.

Materials: Colored pencils and paper

Part A: The Set Model

The set model differs from other fraction models in that the whole is a collection of objects rather than an area or length.

1. Consider the following set of dots.

    a. Circle $\frac{1}{2}$ of the dots in red.　　b. Circle $\frac{1}{3}$ of the dots in blue.

    c. Circle $\frac{3}{4}$ of the dots in green.　　d. Circle $\frac{5}{6}$ of the dots in yellow.

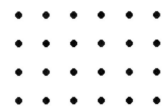

2. Use the following diagram to find the fraction that represents the dots in
    a. The circle as a part of all the dots.
    b. The rectangle as a part of all the dots.
    c. The triangle as a part of all the dots.
    d. Both the triangle and the rectangle as a part of all the dots.
    e. The circle or the triangle as a part of all the dots.
    f. Both the circle and the rectangle as a part of the dots in the rectangle.
    g. The triangle but not the rectangle as a part of the dots in the triangle.

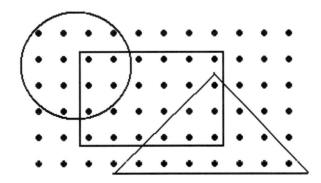

3. Consider the following set of dots.

   a. If the set of dots has a value of $\frac{1}{2}$, how many dots would be equal to 1?

   b. If the set of dots has a value of $\frac{1}{3}$, how many dots would be equal to 1?

   c. If the set of dots has a value of $\frac{1}{4}$, how many dots would be equal to 1?

4. Consider the following set of dots.

   a. If the set of dots has a value of $\frac{1}{3}$, how many dots would be equal to 1?

   b. If the set of dots has a value of $\frac{1}{2}$, how many dots would be equal to $\frac{2}{3}$?

   c. If the set of dots has a value of $\frac{3}{4}$, how many dots would be equal to $\frac{5}{6}$?

5. a. Use the following set of dots to determine whether $\frac{3}{4}$ is less than or greater than $\frac{2}{3}$. Provide a diagram and explanation that illustrate how you used the set of dots to do so.

   b. Can you use the set model to determine whether any fraction is less than or greater than another fraction? If so, what must be true about the number of elements in the set?

Part B: Summary and Connections

1. List four or five real-world situations in which it is appropriate to represent fractions with the set model rather than with the area or length models.

2. Use the set model to design an activity that is appropriate for the elementary grades and develops children's understanding of fractional values.

Exploration 7.4 Student Errors in Understanding the Fractions

Purpose: Students examine common errors children make as they learn to understand fractions and mixed numbers.

Materials: Fraction bars, fraction disks, paper, and pencil

Part A: Errors in Understanding the Meaning of a Fraction

As children work with fractions, it is not unusual for them to make mistakes. Some of their more common errors in thinking follow.

1. Christopher claims that the rectangle on the right represents the larger fractional value because it is a larger region. Is he correct?

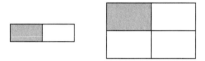

2. Kelli uses the following picture to represent $\frac{1}{3}$. What is wrong with her picture?

3. Nikki says that the following diagram represents $1\frac{1}{2}$. Isaiah says that it represents $\frac{3}{4}$. Is one wrong and the other right? Could they both be correct?

Part B: Errors with Equivalent Fractions

1. Leanne says that it does not make sense that $\frac{2}{3} = \frac{4}{6}$ because 2 does not equal 4 and 3 does not equal 6. Draw a diagram that illustrates how you might use fraction disks to help her understand her misconception.

2. Bella states that $\frac{4}{5} = \frac{4+5}{5+5} = \frac{9}{10}$. Why would Bella think this? How would you correct her thinking?

3. Cornelia converts $\frac{14}{5}$ to a mixed number and gets $4\frac{2}{5}$. Draw a diagram that illustrates how you might use fraction disks to help her understand her mistake.

4. Hunter claims that $-2\frac{1}{2} = -\frac{3}{2}$. Why would Hunter think this? How would you correct his thinking?

Part C: Errors in Ordering Fractions

1. Nadia claims that $\frac{1}{3}$ is smaller than $\frac{1}{4}$ because 3 is smaller than 4. Draw a diagram that illustrates how you might use a number line to correct her thinking.

2. Ying claims that $\frac{4}{8}$ is greater than $\frac{2}{3}$ because 4 is larger than 2. Draw a diagram that illustrates how you might use fraction disks to correct her thinking.

3. Jamal does not think there can be another fraction between $\frac{3}{5}$ and $\frac{4}{5}$ because there is not another whole number between 3 and 4. Draw a diagram that illustrates how you might use fraction bars to correct his thinking.

Part D: Summary and Connections

1. What other mistakes do you think students might make as they come to understand fractions and their properties?

2. Why are manipulatives and other concrete models important for helping students understand their mistakes?

Exploration 7.5 Why is a Common Denominator Necessary?

Purpose: Students explore the need for a common denominator when adding or subtracting fractions.

Materials: Paper and pencil

Part A: Adding and Subtracting Whole Numbers and Other Measures

Although many students can compute sums and differences with fractions, they have trouble understanding why it is necessary to get a common denominator when doing so. To better understand the need for a common denominator, we can use what we know about adding and subtracting whole numbers and measures.

1. a. Find the following sums or differences.
      i. 1 ten + 3 tens   ii. 1 hour + 3 hours   iii. 8 ones − 5 ones
   b. How are the sums and differences alike? How are they different?

2. Consider the standard algorithm for adding 34 + 25.
      a. Why do we add the 4 and 5 and the 3 and 2?
      b. Why do we not add the 3 and 5 or the 4 and 2?

3. a. Find the following sums or differences.
      i. 5 tens + 17 ones   ii. 3 yards − 2 feet   iii. 4 hundreds − 15 tens
   b. What had to be done to find each sum or difference? How does this make them different from the problems in question 1?

4. Consider the standard algorithm for subtracting 45 − 26. What must happen before the ones digits can be subtracted? Why?

5. Write several sentences that summarize what the previous problems indicate about the importance of the units when we add or subtract.

Part B: Adding Fractions with a Common Denominator

In Part A, we learned that only numbers with the same units can be combined in addition. That is, we add ones to ones, tens to tens, hundreds to hundreds, and so on. Likewise, we add minutes to minutes, hours to hours, and days to days. If we want to add numbers with different units, then we must trade or regroup those units first. This idea can be directly applied to adding and subtracting fractions. Consider the following problems.

1. Papa Pete's Pizza allows customers to specify how many equal slices they want their pizza to be cut. This is handy for Bill because he always wants to eat one half of a pizza.

a. Bill could ask Papa Pete to cut his pizza into 2 equal pieces. However, this is not his only option. Give four other ways that Bill could ask Papa Pete to cut his pizza so that he can eat half the pizza without making additional cuts.
b. What observations can you make about the number of slices into which the pizza must be cut so that Bill can always eat $\frac{1}{2}$ of the pizza with no additional cuts?

2. Eating at Papa Pete's Pizza is also convenient for Jane because she always wants to eat one third of a pizza.
a. Jane could ask Papa Pete to cut her pizza into 3 equal slices. However, this is not her only option. Give four other ways that Jane could ask Papa Pete to cut her pizza so that she can eat a third of the pizza without making additional cuts.
b. What observations can you make about the number of slices into which the pizza must be cut so that Jane can always eat $\frac{1}{3}$ of the pizza with no additional cuts?

3. Now suppose Bill and Jane want to order one pizza from Papa Pete's and share it. Bill still wants his half and Jane still wants her third.
a. Is it possible for them to share one pizza? How do you know?
b. $\frac{1}{2} + \frac{1}{3}$ or 1 half + 1 third is the amount of pizza Bill and Jane eat together. What must be done to combine one half and one third? Why?
c. Into how many *equal* slices should Bill and Jane have their pizza cut so that Bill can eat his half and Jane can eat her third without making additional cuts? Give four possibilities.
d. Once Bill and Jane are finished, how much of the pizza is left?

4. Suppose Bill's and Jane's eating habits change so that Bill always eats one fourth of a pizza and Jane always eats one sixth.
a. Is it possible for them to share one pizza? How do you know?
b. $\frac{1}{4} + \frac{1}{6}$ or 1 fourth + 1 sixth is the amount of pizza Bill and Jane eat together. What must be done to combine one fourth and one sixth? Why?
c. Into how many *equal* slices should Bill and Jane ask their pizza to be cut so that Bill can eat his fourth and Jane can eat her sixth without making additional cuts? Give four possibilities.
d. Once Bill and Jane are finished, how much of the pizza is left?

Part C: Summary and Connections

1. Summarize what you have learned about the need for common denominators when adding and subtracting fractions.

2. How can you adapt the pizza problems from Part B into an activity that will help children understand the need for common denominators when subtracting fractions?

Exploration 7.6 Adding and Subtracting Fractions

Purpose: Students use pattern blocks and colored rods to add and subtract fractions.

Materials: Pattern blocks, isometric dot paper, and colored rods

Part A: Adding and Subtracting Fractions with Pattern Blocks

Use pattern blocks to model each sum or difference. In each case, draw a diagram on isometric dot paper illustrating your method. Then write a brief description telling what you did with the pattern blocks and record your work numerically.

1. Find each sum if the hexagon represents 1.

    a. $\frac{1}{3} + \frac{1}{6}$
    b. $\frac{1}{2} + \frac{5}{6}$
    c. $1\frac{1}{2} + \frac{2}{3}$

2. a. Find each difference using the given approach. Let the hexagon represent 1.

    i. $\frac{5}{6} - \frac{1}{3}$ using the take-away approach.
    ii. $\frac{5}{6} - \frac{1}{3}$ using the comparison approach.

    b. What similarities and differences you see in the way you used pattern blocks to compute each difference?

3. Find each sum if ⬡⬡ represents 1.

    a. $\frac{1}{2} + \frac{5}{12}$
    b. $\frac{3}{4} + \frac{2}{3}$

4. Find each difference using the given approach. Let ⬡⬡ represent 1.

    a. $\frac{5}{6} - \frac{1}{3}$ using the take-away approach.

    b. $1\frac{7}{12} - \frac{5}{6}$ using the comparison approach.

5. a. Use pattern blocks to model the answer to $\frac{2}{9} + \frac{2}{3}$.

    b. What shape or arrangement of shapes did you use as 1? How did you determine which arrangement or shape to use?

6. Susie claims $\frac{1}{2} + \frac{1}{4}$ can be modeled using ◇ to represent 1. Do you agree or disagree? Explain your thinking.

Part B: Adding and Subtracting Fractions with Colored Rods

1. a. Use colored rods to find $\frac{1}{2} + \frac{2}{3}$. Draw a diagram to show how you found the sum.

   b. What rod(s) did you use to represent 1? To represent $\frac{1}{2}$? To represent $\frac{2}{3}$?

   c. Why is it important to select the rod(s) to represent 1 as the initial step?

2. Use your method from previous question to compute each sum.

   a. $\frac{1}{3} + \frac{1}{6}$
   b. $\frac{1}{2} + \frac{5}{12}$
   c. $\frac{2}{9} + \frac{2}{3}$
   d. $1\frac{1}{2} + \frac{2}{3}$

3. a. Use colored rods and the take-away approach to find $\frac{5}{6} - \frac{1}{3}$. Draw a diagram to show how you found the difference. What rod(s) did you use to represent 1? To represent $\frac{5}{6}$? To represent $\frac{1}{3}$?

   b. Use colored rods to find $\frac{5}{9} - \frac{1}{3}$ using the missing-addend approach. Draw a diagram to shown how you found the difference. What rod(s) did you use to represent 1? To represent $\frac{5}{9}$? To represent $\frac{1}{3}$?

   c. Are there any similarities or differences in how you used colored rods to compute the differences with the two approaches?

4. Use your methods from the previous question to compute each difference.

   a. $\frac{1}{3} - \frac{1}{6}$
   b. $\frac{5}{8} - \frac{1}{2}$
   c. $\frac{3}{4} - \frac{3}{8}$
   d. $1\frac{7}{12} - \frac{5}{6}$

5. How do you determine which rod(s) represent 1 when adding or subtracting two fractions? What does this rod have to do with a common denominator?

Part C: Summary and Connections

1. a. Which manipulative do you think was the easiest to use when adding and subtracting fractions: pattern blocks or colored rods?
   b. Do you think these manipulatives would help children to make better sense of adding and subtracting fractions? Why or why not?

2. Write a brief summary of what you have learned about adding and subtracting fractions through this exploration.

Exploration 7.7 Fraction Sense with Addition and Subtraction

Purpose: Students develop their number sense by adding and subtracting fractions.

Materials: Paper and pencil

Part A: Developing Fraction Sense with Addition and Subtraction

1. Using only estimates, insert <, >, or = in each circle to make the expression true.

   a. $\frac{9}{16} + \frac{13}{24} \bigcirc 1$
   b. $\frac{7}{16} + \frac{23}{24} \bigcirc 1\frac{1}{2}$
   c. $\frac{1}{16} + \frac{7}{24} \bigcirc \frac{1}{4}$
   d. $\frac{6}{24} + \frac{1}{9} \bigcirc \frac{1}{2}$

2. Using only estimates, insert <, >, or = in each circle to make the expression true.

   a. $\frac{7}{9} - \frac{1}{11} \bigcirc \frac{1}{2}$
   b. $\frac{5}{9} - \frac{5}{11} \bigcirc 0$
   c. $\frac{10}{11} - \frac{4}{9} \bigcirc \frac{2}{3}$
   d. $\frac{9}{11} - \frac{5}{9} \bigcirc \frac{1}{2}$

3. Suppose p and q are rational numbers such that $0 < p < q < 1$. Determine whether each statement is true, false, or cannot be determined.
   a. $p + q > 0$
   b. $p + q < 1$
   c. $p + q < q$
   d. $p + q > q$
   e. $q - p < 0$
   f. $q - p > \frac{1}{2}$
   g. $q - p > p$
   h. $q - p < p$

4. If possible, find two positive rational numbers p and q, such that $p < q$ and
   a. $p + q < 1$
   b. $p + q > 1$
   c. $q - p < p$
   d. $p < q - p < q$
   e. $q - p < 0$
   f. $q - p > q$

Part B: Summary and Connections

1. Why is estimation a useful strategy for answering the questions in Part A?

2. How could you use a number line to solve the first two questions in Part A?

3. a. How were the questions in Part A valuable in developing your number sense with rational number addition and subtraction?
   b. Do you think questions like those in Part A would be reasonable to use in a fifth or sixth grade classroom? If not, how would you revise them so that they are?

Exploration 7.8 Student Errors in Adding and Subtracting Fractions

Purpose: Students examine common errors that children make when adding and subtracting fractions and mixed numbers.

Materials: Paper and pencil

Part A: Children's Errors with Adding Fractions and Mixed Numbers

When children learn to add fractions and mixed numbers, they often make mistakes. Part of your task as a teacher will be to diagnose their errors and give them feedback for corrections. Examine the problems solved by each student. In each case, identify the error in the student's thinking and then explain how you might correct it.

1. David:

$$\frac{3}{5} + \frac{2}{3} = \frac{5}{8} \qquad \frac{4}{3} + \frac{5}{6} = \frac{9}{9} = 1$$

2. Mikaela:

$$\frac{2}{5} + \frac{1}{2} = \frac{2}{10} + \frac{1}{10} = \frac{3}{10} \qquad \frac{1}{3} + \frac{3}{4} = \frac{1}{12} + \frac{3}{12} = \frac{4}{12} = \frac{1}{3}$$

3. Sophie:

$$2\frac{1}{2} + 3\frac{1}{2} = 5\frac{1}{2} \qquad 3\frac{1}{4} + 7\frac{1}{4} = 10\frac{1}{4}$$

4. James:

$$2\frac{3}{4} + 1\frac{3}{4} = 3\frac{6}{4} = 3\frac{3}{2} = 6\frac{1}{2} \qquad 4\frac{5}{8} + 2\frac{7}{8} = 6\frac{12}{8} = 6\frac{3}{2} = 9\frac{1}{2}$$

Part B: Children's Errors with Subtracting Fractions and Mixed Numbers

Children also have trouble when subtracting fractions and mixed numbers. Examine the problems solved by each student. In each case, identify the error in the student's thinking and then explain how you might correct it.

1. Trevor:

$$\frac{7}{9} - \frac{3}{6} = \frac{7}{9} - \frac{6}{9} = \frac{1}{9} \qquad \frac{5}{8} - \frac{7}{12} = \frac{9}{12} - \frac{7}{12} = \frac{2}{12} = \frac{1}{6}$$

2. Kristin:

$$8\frac{1}{6} - 3\frac{5}{6} = 5\frac{4}{6} = 5\frac{2}{3} \qquad 9\frac{2}{5} - 6\frac{3}{5} = 3\frac{1}{5}$$

3. Jonathan:

$$9 - 4\frac{2}{3} = 5\frac{2}{3} \qquad 11 - 6\frac{3}{8} = 5\frac{3}{8}$$

4. Javon:

$$6\frac{1}{5} - 2\frac{3}{5} = 5\frac{11}{5} - 2\frac{3}{5} = 3\frac{8}{5} = 4\frac{3}{5} \qquad 5\frac{3}{7} - 1\frac{5}{7} = 4\frac{13}{7} - 1\frac{5}{7} = 3\frac{8}{7} = 4\frac{1}{7}$$

5. Roopa:

$$7\frac{1}{4} - 2\frac{1}{2} = 7\frac{5}{4} - 2\frac{2}{4} = 5\frac{3}{4} \qquad 9\frac{3}{8} - 4\frac{3}{4} = 9\frac{11}{8} - 4\frac{6}{8} = 5\frac{5}{8}$$

Part C: Summary and Connections

1. Why do you think that adding and subtracting fractions and mixed numbers is difficult for many students to master? As a teacher, what steps can you take to help students learn these skills more easily?

2. a. Give at least three real-world situations in which you have added or subtracted fractions or mixed numbers.
   b. How might using a context help students to make sense of adding and subtracting fractions and mixed numbers?

152   Chapter 7 Fractions and the Rational Numbers

Exploration 7.9 The Meaning of Fraction Multiplication

Purpose: Students use concrete models to make sense of fraction multiplication.

Materials: Paper and pencil

Part A: The Product of a Whole Number and a Fraction

We can gain a better understanding of fraction multiplication by revisiting the approaches we used to compute whole number products. In general, we used two approaches with whole numbers: the **Cartesian product approach** and the **repeated-addition approach**. To determine whether these approaches can be used with fractions or will need to be adapted, we first consider what happens when we use them to compute the product of a whole number and a fraction.

1. In the Cartesian-product approach, we pair the elements from two sets and then count the total number of pairs. Is it possible to use the Cartesian product approach to compute $6 \cdot \frac{1}{2}$? If so, explain how you can use sets to represent the Cartesian product. If not, why?

2. In the repeated-addition approach, the product $a \cdot b$ is interpreted as the sum of b added to itself a times, or $a \cdot b = \underbrace{b + b + \ldots + b}_{a \text{ addends}}$. Use a repeated sum to compute each product.

    a. $6 \cdot \frac{1}{2}$ b. $4 \cdot \frac{1}{8}$ c. $10 \cdot \frac{1}{5}$

3. a. It is more difficult to interpret the product $\frac{1}{3} \cdot 9$ as a repeated sum. Why?

   b. How could you use repeated addition and the commutative property of multiplication to find this product?

   c. Based on your answer to part (b), what is another way to interpret $\frac{1}{3} \cdot 9$?

   d. Use your answer to part (c) to interpret and compute the each product.

    i. $\frac{1}{2} \cdot 8$ ii. $\frac{1}{7} \cdot 14$ iii. $\frac{1}{6} \cdot 18$

Part B: The Product of Two Fractions

In Part A, we learned that one way to interpret the product $\frac{1}{3} \cdot 9$ is to take $\frac{1}{3}$ of 9. We can use the same idea to interpret the product of two fractions. For instance, we can interpret

$\frac{1}{2} \cdot \frac{1}{3}$ as "Take one half of one third," and $\frac{1}{3} \cdot \frac{3}{4}$ as "Take one third of three fourths." To find the value of these products, we can use fraction bars. For instance, to model $\frac{1}{2} \cdot \frac{1}{3}$, take $\frac{1}{3}$ of a fraction bar, divide it into two equal pieces, and shade one of them. We then determine the value of this piece relative to the bar that represents 1. Because it is $\frac{1}{6}$ of this bar, $\frac{1}{2} \cdot \frac{1}{3} = \frac{1}{6}$.

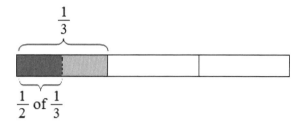

1. Give a verbal interpretation of each product, and then draw a diagram that illustrates how to use fraction bars to find its value.

   a. $\frac{1}{4} \cdot \frac{1}{3}$ b. $\frac{2}{3} \cdot \frac{2}{5}$ c. $\frac{1}{5} \cdot \frac{3}{4}$ d. $\frac{1}{8} \cdot \frac{1}{2}$

2. Examine your answers to the products in the previous problem.
   a. Do you see a relationship between the numerators of the factors and the numerator of the product?
   b. Do you see a relationship between the denominators of the factors and the denominator of the product?
   c. Based on your observations, suggest an algorithm that can be used to compute the product of two fractions.
   d. Use your algorithm from part (c) to compute each product.

   i. $\frac{4}{7} \cdot \frac{5}{9}$ ii. $\frac{2}{9} \cdot \frac{5}{8}$ iii. $\frac{6}{11} \cdot \frac{3}{4}$ iv. $\frac{9}{10} \cdot \frac{7}{8}$

3. Devise a way to use the area of a rectangle to model the product of two fractions. Draw a diagram that illustrates how to use your method to compute the products in Question 1.

Part C: Summary and Connections

1. Write a story problem that can be used to solve the product $\frac{1}{4} \cdot \frac{1}{3}$. Do you think placing the product of two fractions in a context helps to make sense of the product?

2. How is multiplying 4 by $5\frac{3}{4}$ similar to multiplying 4 by 57?

Exploration 7.10 The Meaning of Fraction Division

Purpose: Students use the context of a problem to make sense of why dividing by a fraction is equivalent to multiplying by its reciprocal.

Materials: Paper and pencil

Part A: Dividing Four Cakes

When most people are asked to divide two fractions, they quickly apply the "invert and multiply" rule. Unfortunately, most people use this rule with little to no understanding of why it works. The following sequence of tasks is designed to help you make sense of this procedure.

1. Wanda really likes cake. She has 4 cakes and decides that one serving should be 2 cakes.
   a. How many servings does Wanda have? Draw a picture to solve the problem.
   b. Why can the quotient 4 ÷ 2 be used to solve this problem?
   c. Did you use repeated subtraction or partitioning to compute the quotient? Explain your thinking.

2. a. Look back at your visual model. Did you draw something that looks like the following picture? If not, how does your representation differ from the one shown?

   b. Discuss with a group of your peers how to make sense of the picture and how the picture can be used to answer the previous question.

3. Reconsider the picture in the previous question.
   a. How many servings do the first two cakes represent? The second two?
   b. If two cakes represent one serving, then each cake represents how many servings?
   c. How can you then count the number of servings using the picture? How is your counting related to repeated addition?
   d. This implies 4 ÷ 2 is equivalent to what product?

Part B: Dividing Five Cakes

1. Suppose Wanda has 5 cakes and decides that one serving should be 2 cakes.
   a. How many servings does Wanda have so that no cake remains? Draw a picture to solve the problem.
   b. Why can the quotient 5 ÷ 2 be used to solve this problem?

c. Did you use repeated subtraction or partitioning to compute the quotient? Explain your thinking.

2. a. Look back at your visual model. Did you draw something that looks like the following picture? If not, how does your representation differ from the one shown?

b. Discuss with a group of your peers how to make sense of the picture and how the picture can be used to answer the previous question.

3. Reconsider the picture in the previous question.
   a. How many servings do the first two cakes represent? The second two? The last cake?
   b. How can you then count the number of servings using the picture? How is your counting related to repeated addition?
   c. This implies 5 ÷ 2 is equivalent to what product?

Part C: Dividing Four Cakes by a Fractional Amount

1. Now suppose Wanda has 4 cakes and decides that one serving should be $\frac{3}{5}$ of a cake.

   a. How many servings does Wanda have? Draw a picture to solve the problem.
   b. Why can the quotient $4 \div \frac{3}{5}$ be used to solve this problem?

2. a. Look back at your visual model. Did you draw something that looks like the following picture? If not, how does your representation differ from the one shown?

   | 1 | 1 | 1 | 2 | 2 | | 2 | 3 | 3 | 3 | 4 | | 4 | 4 | 5 | 5 | 5 | | 6 | 6 | 6 | |

   b. Discuss with a group of your peers how to make sense of the picture and how the picture can be used to answer the previous question.

3. Reconsider the picture in the previous question.
   a. How many servings does each cake contain?
   b. How can you count the number of servings using the picture? How is your counting related to repeated addition?
   c. This implies $4 \div \frac{3}{5}$ is equivalent to what product?

Part D: Dividing a Fractional Amount by a Fractional Amount

1. Finally, suppose Wanda has $3\frac{1}{2}$ cakes and decides that one serving should be $\frac{1}{4}$ of a cake.
    a. How many servings does Wanda have? Draw a picture to solve the problem.
    b. Why can the quotient $3\frac{1}{2} \div \frac{1}{4}$ be used to solve this problem?

2. a. Look back at your visual model. Did you draw something that looks like the following picture? If not, how does your representation differ from the one shown?

$$\boxed{1|2|3|4} \quad \boxed{5|6|7|8} \quad \boxed{9|10|11|12} \quad \boxed{13|14}$$

   b. Discuss with a group of your peers how to make sense of the picture and how the picture can be used to answer the previous question.

3. Reconsider the picture in the previous question.
    a. How many servings does each cake contain?
    b. How can you count the number of servings using the picture?
    c. This implies $3\frac{1}{2} \div \frac{1}{4}$ is equivalent to what product?

Part E: Summary and Connections

1. Write several sentences in which you explain your understanding of why the "invert and multiply" rule works?

2. Write a word problem that uses the calculation $\frac{1}{2} \div \frac{1}{4}$, and use it to explain why $\frac{1}{2} \cdot 4$ can be used to solve the problem.

3. Write a word problem that uses the calculation $2\frac{5}{6} \div \frac{2}{3}$ and use it to explain why $2\frac{5}{6} \cdot \frac{3}{2}$ can be used to solve the problem.

Exploration 7.11 Multiplying and Dividing Fractions

Purpose: Students use pattern blocks and colored rods to multiply and divide fractions.

Materials: Pattern blocks, isometric dot paper, and colored rods

Part A: Multiplying Fractions and Mixed Numbers with Pattern Blocks

Use pattern blocks to model each product. In each case, draw a diagram on isometric dot paper illustrating your method. Then write a brief description of what you did with the pattern blocks and record your work numerically.

1. Compute each product involving a whole number and a fraction.

   a. $4 \cdot \dfrac{2}{3}$ b. $5 \cdot \dfrac{1}{6}$ c. $2 \cdot \dfrac{3}{4}$

2. Compute each product involving a fraction and a fraction.

   a. $\dfrac{1}{2} \cdot \dfrac{1}{3}$ b. $\dfrac{3}{4} \cdot \dfrac{2}{3}$ c. $\dfrac{1}{2} \cdot \dfrac{3}{4}$ d. $\dfrac{1}{4} \cdot \dfrac{3}{12}$

3. Compute each product involving two mixed numbers.

   a. $2\dfrac{1}{2} \cdot \dfrac{1}{2}$ b. $1\dfrac{1}{4} \cdot \dfrac{2}{3}$ c. $\dfrac{2}{3} \cdot 3\dfrac{1}{2}$ d. $1\dfrac{1}{3} \cdot 2\dfrac{1}{2}$

4. a. How did your approach to multiplication change from one set of products to another?
   b. How did you decide which block(s) would represent 1? What about each factor in the product?

Part B: Multiplying Fractions and Mixed Numbers with Colored Rods

Use colored rods to model each product. Draw a diagram illustrating your method and write a brief description of what you did with the rods. Record your work numerically.

1. Compute each product involving a whole number and a fraction.

   a. $6 \cdot \dfrac{1}{2}$ b. $3 \cdot \dfrac{1}{5}$ c. $3 \cdot \dfrac{2}{3}$ d. $2 \cdot \dfrac{3}{4}$

2. Compute each product involving a fraction and a fraction.

   a. $\dfrac{1}{2} \cdot \dfrac{1}{3}$ b. $\dfrac{3}{4} \cdot \dfrac{1}{2}$ c. $\dfrac{5}{6} \cdot \dfrac{2}{3}$ d. $\dfrac{1}{3} \cdot \dfrac{3}{5}$

158   Chapter 7 Fractions and the Rational Numbers

3. Compute each product involving mixed numbers.

   a. $3\frac{1}{2} \cdot \frac{2}{5}$
   b. $1\frac{3}{4} \cdot \frac{1}{6}$
   c. $3\frac{1}{2} \cdot 2\frac{1}{4}$
   d. $1\frac{1}{3} \cdot 2\frac{1}{2}$

4. a. How did your approach to multiplication change from one set of products to another?
   b. How did you decide which rod(s) would represent 1? What about each factor in the product?

Part C: Dividing Fractions and Mixed Numbers with Pattern Blocks

Use pattern blocks to model each quotient. In each case, draw a diagram on isometric dot paper illustrating your method. Then write a brief description of what you did and record your work numerically. It may be useful to put the problem in a context.

1. Use pattern blocks and repeated subtraction to find each quotient.

   a. $2 \div \frac{1}{3}$
   b. $3 \div \frac{2}{3}$
   c. $\frac{1}{2} \div \frac{1}{3}$
   d. $1\frac{1}{4} \div \frac{1}{6}$

2. a. Use pattern blocks and partitioning to find each quotient.

   i. $2\frac{1}{2} \div 5$
   ii. $\frac{1}{2} \div 3$
   iii. $\frac{2}{3} \div 2$

   b. Is it possible to use pattern blocks and partitioning to solve $3 \div \frac{1}{2}$? Why or why not?

Part D: Dividing Fractions and Mixed Numbers with Colored Rods

Use colored rods to model each quotient. Draw a diagram illustrating your method and write a brief description of what you did with the rods. Record your work numerically. Again, it may be useful to put each problem in a context.

1. Use colored rods and repeated subtraction to find each quotient.

   a. $2 \div \frac{1}{6}$
   b. $3 \div \frac{2}{5}$
   c. $\frac{1}{2} \div \frac{3}{8}$
   d. $1\frac{2}{3} \div \frac{1}{6}$

2. a. Use colored rods and partitioning to find each quotient.

   i. $\frac{1}{3} \div 3$
   ii. $\frac{3}{4} \div 3$
   iii. $2\frac{2}{3} \div 8$

   b. Is it possible to use colored rods and partitioning to solve $4 \div \frac{1}{4}$? Why or why not?

Part E: Summary and Connections

1. In general, what limitations are there to using pattern blocks and colored rods to find products and quotients involving fractions?

2. a. Which manipulative do you think was the easiest to use when multiplying and dividing: pattern blocks or colored rods?

   b. Do you think these manipulatives can help children to make better sense of multiplying and dividing fractions? If so, how?

3. Write a brief summary of what you have learned about multiplying and dividing fractions through this exploration?

160   Chapter 7 Fractions and the Rational Numbers

Exploration 7.12 Fraction Sense with Multiplication and Division

Purpose: Students develop their number sense by multiplying and dividing fractions.

Materials: Paper and pencil

Part A: Developing Fraction Sense with Multiplication and Division

1. Using only estimates, insert <, >, or = in each circle to make the expression true.

   a. $\frac{8}{19} \cdot \frac{12}{25} \bigcirc \frac{1}{2}$ b. $\frac{18}{19} \cdot \frac{24}{25} \bigcirc 1$ c. $\frac{1}{19} \cdot \frac{1}{25} \bigcirc 0$ d. $\frac{18}{19} \cdot \frac{24}{25} \bigcirc \frac{24}{25}$

2. Using only estimates, insert <, >, or = in each circle to make the expression true.

   a. $\frac{5}{9} \div \frac{4}{10} \bigcirc 1$ b. $\frac{9}{10} \div \frac{8}{9} \bigcirc 1$ c. $\frac{5}{9} \div \frac{13}{10} \bigcirc \frac{5}{9}$ d. $\frac{5}{9} \div \frac{9}{10} \bigcirc \frac{5}{9}$

3. Suppose p and q are rational numbers such that $0 < p < q < 1$. Determine whether each statement is true, false, or cannot be determined.

   a. $p \cdot q < 0$ b. $p \cdot q < \frac{1}{2}$ c. $p \cdot q > 1$ d. $p \cdot q > q$

   e. $q \div p < 0$ f. $q \div p > 1$ g. $q \div p > p$ h. $q \div p < q$

4. If possible, find two positive rational numbers p and q, such that $p < q$ and
   a. $p \cdot q < 1$ b. $p \cdot q > 1$ c. $p \cdot q < q$
   d. $q \div p < p$ e. $q \div p > q$ f. $p < q \div p < q$

Part B: Summary and Connections

1. Why is estimation a useful strategy for answering the questions in Part A?

2. a. How were the questions in Part A valuable in developing your number sense with rational number multiplication and division?
   b. Do you think questions like those in Part A would be reasonable to use in a fifth or sixth grade classroom? If not, how would you revise them so that they are?

Exploration 7.13 Computing with a Fraction Calculator

Purpose: Students learn to use a fraction calculator to simplify and perform operations with fractions.

Materials: Fraction calculator

Part A: Simplifying Fractions

Take a moment to familiarize yourself with the features on your calculator which allow you to simplify fractions. Use these features to answer each question.

1. Write the sequence of keys you would press to simplify $\frac{18}{45}$.

2. Simplify each fraction. Express your answers as fractions not as mixed numbers.

   a. $\frac{42}{56}$ b. $-\frac{88}{132}$ c. $\frac{195}{91}$ d. $-\frac{147}{168}$

Part B: Conversions Between Fractions and Mixed Numbers

Most fraction calculators can also make conversions between fractions and mixed numbers. Familiarize yourself with these features on your calculator and then use them to answer each question.

1. a. Write the sequence of keys you would press to convert $\frac{13}{2}$ to a mixed number.

   b. Write the sequence of keys you would press to convert $1\frac{7}{8}$ to an improper fraction.

2. Use a calculator to convert each fraction to a mixed number with a simplified fractional part.

   a. $\frac{39}{14}$ b. $\frac{135}{28}$ c. $-\frac{55}{12}$ d. $-\frac{34}{8}$

3. Use a calculator to convert each mixed number to an improper fraction.

   a. $7\frac{15}{22}$ b. $6\frac{4}{11}$ c. $-7\frac{4}{9}$ d. $-8\frac{11}{19}$

Part C: Operations on Fractions and Mixed Numbers

Computations on fractions and mixed numbers can also be done with a fraction calculator.

1. Write the sequence of keys you would press on your calculator to compute each problem and get an answer in a simplified, fractional form.

   a. $\dfrac{3}{10} + \dfrac{2}{7}$ b. $\dfrac{56}{35} \cdot \dfrac{25}{64}$.

2. Compute each problem. Be sure all answers are in fractional form and simplified.

   a. $\dfrac{7}{13} + \dfrac{2}{5}$ b. $\dfrac{4}{5} - \dfrac{1}{12}$ c. $\dfrac{15}{32} \cdot \dfrac{-12}{23}$ d. $\dfrac{16}{15} \div \dfrac{24}{25}$

3. Write the sequence of keys you press on your calculator to compute each problem and get an answer as a mixed number with the fractional part in simplest form.

   a. $7\dfrac{2}{3} - 1\dfrac{1}{7}$ b. $2\dfrac{2}{3} \div 2\dfrac{1}{6}$

4. Compute each problem. Be sure all answers are in mixed number form with a simplified fractional part.

   a. $3\dfrac{1}{10} + 7\dfrac{6}{7}$ b. $8\dfrac{3}{4} - 4\dfrac{4}{5}$ c. $1\dfrac{4}{5} \times 1\dfrac{3}{5}$ d. $3\dfrac{4}{7} \div -1\dfrac{3}{7}$

5. Calculate each problem. Be sure all answers are simplified.

   a. $5\dfrac{5}{7} \times \dfrac{5}{12} - 6\dfrac{2}{9} \div \dfrac{5}{18}$ b. $\dfrac{20}{6} \times \left[\left(\dfrac{16}{5} + 2\dfrac{1}{7} \right) \div 6\dfrac{1}{3} \right]$

Part D: Summary and Connections

1. Do you think that a fraction calculator should play a role in teaching middle grades mathematics? Why or why not?

2. The main objective of this exploration was for students to learn how to use a fraction calculator. In this case, the problems were not placed in a context. Do you think there are advantages or disadvantages to teaching students how to use technology through contextual situations? Explain your thinking.

Exploration 7.14 Student Errors in Multiplying and Dividing Fractions

Purpose: Students examine common errors that children make when multiplying and dividing fraction and mixed numbers.

Materials: Paper and pencil

Part A: Children's Errors with Multiplying Fractions and Mixed Numbers

When children learn to multiply fractions and mixed numbers, they often make mistakes. Part of your task as a teacher will be to diagnose their errors and give them feedback for corrections. Examine the problems solved by each student. In each case, identify the error in the student's thinking and then explain how you might correct it.

1. Tyson:

$$\frac{2}{7} \cdot \frac{3}{7} = \frac{6}{7} \qquad \frac{4}{5} \cdot \frac{3}{5} = \frac{12}{5}$$

2. Josh:

$$\frac{2}{3} \cdot \frac{4}{5} = \frac{10}{12} \qquad \frac{7}{8} \cdot \frac{3}{4} = \frac{28}{24} = \frac{7}{6}$$

3. Katie:

$$2 \cdot \frac{3}{5} = \frac{6}{10} \qquad 4 \cdot \frac{1}{3} = \frac{4}{12} = \frac{1}{3}$$

4. Hunter:

$$3\frac{1}{2} \cdot 2\frac{1}{3} = 6\frac{1}{6} \qquad 5\frac{3}{4} \cdot 5\frac{5}{8} = 25\frac{15}{32}$$

5. Kristin:

$$4 \cdot 5\frac{3}{4} = 20\frac{3}{4} \qquad 3 \cdot 2\frac{5}{7} = 6\frac{5}{7}$$

Part B: Children's Errors with Dividing Fractions and Mixed Numbers

Children also have trouble when dividing fractions and mixed numbers. Examine the problems solved by each student. In each case, identify the error in the student's thinking and then explain how you might correct it.

1. Kenny:

$$\frac{2}{3} \div \frac{4}{5} = \frac{2}{3} \cdot \frac{4}{5} = \frac{8}{15} \qquad \frac{4}{9} \div \frac{3}{7} = \frac{4}{9} \cdot \frac{3}{7} = \frac{12}{63} = \frac{4}{21}$$

2. Rayna:

$$\frac{1}{8} \div 2 = \frac{1}{4} \qquad \frac{5}{9} \div 3 = \frac{5}{3}$$

3. Javai:

$$12 \div \frac{1}{6} = 2 \qquad 15 \div \frac{1}{3} = 5$$

4. Latasha:

$$6\frac{5}{8} \div 3 = 2\frac{5}{8} \qquad 10\frac{4}{9} \div 2 = 5\frac{4}{9}$$

Part C: Summary and Connections

1. Why do you think that multiplying and dividing fractions and mixed numbers can be difficult for many students to master? As a teacher, what steps can you take to help students learn these skills more easily?

2. a. Give two or three real-world situations in which you have multiplied or divided fractions or mixed numbers.
   b. How might using a context help students to make better sense of adding and subtracting fractions and mixed numbers?

Chapter 8
Decimals, Real Numbers, and Proportional Reasoning

The decimals provide another way to represent part-to-whole relationships in a way that is consistent with the ten-to-one ratio of the decimal system. Children first encounter decimal numbers in first and second grade as they study dollar and cents notation for money. Although these early experiences are limited, children do learn to interpret and use different decimal representations. In the third, fourth, and fifth grades, students work in earnest with the decimals. They learn how to order and compute with them by extending what they have learned from whole-number computations.

Fractions and decimals play an important role in **proportional reasoning**, which is often considered the pinnacle of arithmetic. Proportional reasoning refers to making comparisons between quantities and understanding the context in which the comparison takes place. Not only is it practical, but it also brings together many mathematical ideas that students have already learned. Proportional reasoning is limited in the elementary grades because students in those grades have not developed the mathematical knowledge needed to work with it in earnest. However, once children can compute with fractions and decimals, they can proceed through a more formal progression of proportional reasoning. This often starts by developing the ideas of ratios, proportions, and percents and then moves to applying them in solve real-world situations.

The explorations in Chapter 8 use a variety of representations to investigate the fundamental ideas associated with decimals, decimal operations, and proportional reasoning. Specifically,

- Explorations 8.1 - 8.3 consider the fundamental ideas of the decimals and the different ways to represent them.
- Explorations 8.4 - 8.7 focus on decimal operations and their connection to whole-number and rational number operations.
- Explorations 8.8 - 8.11 consider proportional reasoning and the different tools that can be used to solve proportion problems.
- Explorations 8.12 - 8.13 apply proportional reasoning to solve problems with percents.

Exploration 8.1 Base-Ten Blocks and Decimal Numbers

Purpose: Students use base-ten blocks to represent decimal numbers in different ways.

Materials: Base-ten blocks and a place value mat

Part A: Representing Decimals with Base-Ten Blocks

Base-ten blocks can be used to represent decimal numbers by choosing any shape to represent 1. Once the shape for 1 is chosen, the value of the other shapes is determined. For instance, if the flat represents 1, then the large cube has a value of 10, the long has a value of 0.1, and the small cube has a value of 0.01.

1. Suppose the long represents 1. What is the value of the
    a. Large cube?   b. Flat?   c. Small cube?

2. Suppose the large cube represents 1. What is the value of the
    a. Flat?   b. Long?   c. Small cube?

3. Suppose the flat represents 1. What decimal number does each set of blocks represent?
    a. 3 flats   b. 3 large cubes, 3 flats, 1 long, and 7 small cubes
    c. 2 large cubes, and 6 longs   d. 2 large cubes, 4 flats, and 5 small cubes

4. Suppose the large cube represents 1. Use a minimal set of base-ten blocks to represent each decimal number and then draw a diagram that shows the blocks you used.
    a. 0.25   b. 0.048   c. 3.47   d. 1.567

5. Suppose a decimal number is represented with 2 large cubes, 1 flat, and 5 small cubes. What is the decimal if the
    a. Small cube represents 1?   b. Long represents 1?
    c. Flat represents 1?   d. Large cube represents 1?

6. Why is it important to identify the block that represents a value of 1 when representing decimal numbers with base-ten blocks?

Part B: Decomposing and Composing Decimal Numbers in Different Ways

Like whole numbers, decimal numbers can be decomposed in different ways by making exchanges between groups of ten.

1. Use a place value mat and base-ten blocks to represent each decimal number in two ways that are different from the minimal set. Be sure to identify the block that represents a

value of 1. When you are finished, draw a sketch of the base-ten blocks on the mat. Then complete the statements summarizing the groups of ten used to form the number.

a. 3.47 = ____ hundreds + ____ tens + ____ ones + ____ tenths + ____ hundredths

3.47 = ____ hundreds + ____ tens + ____ ones + ____ tenths + ____ hundredths

b. 423.6 = ____ hundreds + ____ tens + ____ ones + ____ tenths + ____ hundredths

423.6 = ____ hundreds + ____ tens + ____ ones + ____ tenths + ____ hundredths

2. Represent the number 13.07 in four different ways by filling in the blanks. Use exchanges rather than calculations to obtain your answers

13.07 = _____ tens + _____ ones + _____ tenths + _____ hundredths

13.07 = _____ tens + _____ ones + _____ tenths + _____ hundredths

13.07 = _____ tens + _____ ones + _____ tenths + _____ hundredths

13.07 = _____ tens + _____ ones + _____ tenths + _____ hundredths

3. Make exchanges to rewrite each decimal number using the fewest number of tens, ones, tenths, hundredths, and thousandths. Finally write the decimal number.

a. 4 ones + 20 tenths + 30 hundredths

= _____ tens + _____ ones + _____ tenths + _____ hundredths + _____ thousandths

= _____

b. 5 tens + 12 ones + 43 tenths + 0 hundredths + 32 thousandths

= _____ tens + _____ ones + _____ tenths + _____ hundredths + _____ thousandths

= _____

Part C: Summary and Connections

1. Do you think using base-ten blocks to represent whole numbers and decimal numbers would be confusing to students? Why or why not?

2. If you did Exploration 3.4, you might notice that strategies used in that exploration and in this one are similar. How so?

Exploration 8.2 Modeling Decimal Numbers

Purpose: Students represent decimals using a variety of models.

Materials: Decimal grid paper

Part A: Verbal and Numerical Representations

A decimal number can be viewed as a notational shortcut for a **decimal fraction**, which is a fraction that has a power of ten in the denominator. To write the decimal number, take the numerator of the fraction and place it in the appropriate place value to the right of the decimal point. For instance, $\frac{6}{10}$ can be written as 0.6 and $\frac{53}{100}$ can be written as 0.53.

1. Write each decimal fraction as a decimal number.
   a. $\frac{3}{10}$ b. $\frac{47}{100}$ c. $\frac{301}{1000}$ d. $\frac{45}{10}$ e. $\frac{5,673}{10,000}$

2. Write each decimal number as a decimal fraction.
   a. 0.4   b. 0.05   c. 0.076   d. 1.03   e. 14.2

3. Write the name of each decimal number.
   a. 0.5   b. 0.43   c. 4.02   d. 74.083

4. Write the decimal number given by each decimal name.
   a. Seven tenths   b. Sixty-five hundredths.
   c. Seven and three hundredths.   d. Fifty-one and fourteen thousandths.

5. Why do we use the same terminology to name decimal fractions and decimal numbers?

Part B: Decimal Grid Paper

Decimal grid paper is a version of the area model that allows us to work with decimal places. Once we determine what represents 1, we can then represent a particular decimal number by shading in the appropriate number of squares. In questions 1 and 2, let the following square have a value of 1.

1. What decimal number is represented in each diagram?

   a.

   b.

   c.

   d.

2. Use decimal grid paper to represent each decimal number by shading.
   a. 0.31   b. 0.08   c. 0.7   d. 0.435

3. What is the smallest, possible decimal that can be represented on decimal grid paper? What must square must represent 1 in this case?

4. a. How can you use decimal grid paper to show that 0.4 > 0.23?
   b. How can you use decimal grid paper to show that 0.126 < 0.128?
   c. Does the square that represents a value of 1 change in the previous questions?

Part C: Money

In the elementary classroom, students are often introduced to decimal numbers while they learn about money.

1. If a dollar always has a value of 1, what is the value of a dime? A penny?

2. What decimals are represented by the following money amounts?

   a.

   b.

   c.

   d.

2. Draw a diagram that illustrates the money amounts needed to represent each decimal.
   a. 0.7   b. 0.14   c. 2.45   d. 14.06

3. What are the limitations of representing decimal numbers with money? Why do the limitations exist?

Part D: Number Line

We can also locate decimal numbers on the number line.

1. Locate the points on the number line associated with following decimal numbers. Label each point with the appropriate capital letter.
   A. 0.6   B. -0.25   C. 1.2   D. – 1.75   E. 0.9

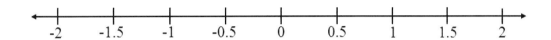

2. Estimate the location of each named point on the number line to the nearest tenth.

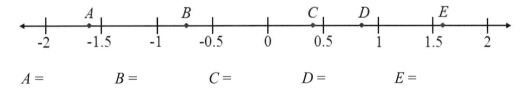

A =   B =   C =   D =   E =

Part E: Summary and Connections

1. Which of the models in this exploration is the best way to represent decimals? The worst?

2. Why is money one of the most common ways to introduce children to the decimals?

3. Because a decimal number is a notational shortcut for a decimal fraction, some fraction models can be used to represent decimal numbers.
   a. What decimal numbers, if any, can be easily represented with fraction disks? With fraction bars?
   b. What are the limitations in using fraction disks and bars to represent decimals?

Exploration 8.3 Understanding Repeating Decimals and Irrational Numbers

Purpose: Students use different representations to understand the infinite nature of repeating decimals and irrational numbers.

Materials: Internet access, calculator, paper, and pencil

Part A: Understanding Repeating Decimals

A **repeating decimal** is a decimal number that has a block of digits that continuously repeats. In other words, they cannot be represented with a single decimal fraction. To get a better understanding of repeating decimals, we can represent them with fraction bars. For instance, consider $\frac{1}{3}$. Figure 1 shows that $\frac{1}{3}$ is equivalent to three tenths, three hundredths, three thousandths, and so on. Each time we make a comparison, there is always a little piece remaining that is equivalent to 3 of the next smallest decimal fraction plus a little bit more. Consequently, $\frac{1}{3} = \frac{3}{10} + \frac{3}{100} + \frac{3}{1,000} + ... = 0.3333...$.

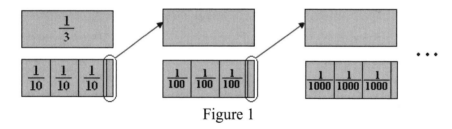

Figure 1

1. Draw a diagram like Figure 1 that uses fraction bars to represent each repeating decimal.
   a. 0.222... b. 0.1333... c. 0.212121...

2. The repeating decimal representation of a fraction can be found by dividing the numerator by the denominator. Use division to find the repeating decimal for each fraction. Explain what happens in the division that causes the decimal to repeat.
   a. $\frac{1}{6}$ b. $\frac{4}{9}$ c. $\frac{6}{11}$ d. $\frac{4}{7}$

3. Reconsider Figure 1.
   a. How is division represented in Figure 1? Which approach to division is used: repeated subtraction, a missing-factor, or partitioning?
   b. How could you combine Figure 1 with division to explain why some fractions have repeating decimal representations?

Part B: Understanding Irrational Numbers

Irrational numbers are decimal numbers that do not terminate and do not repeat. In other words, irrational numbers, like repeating decimals, cannot be written as a single decimal fraction. For instance, suppose 0.2451... represents the first four digits of an irrational number. Figure 2 uses fraction bars to show that 0.2451... is equivalent to two tenths, four hundredths, and so on. As with repeating decimals, each time we use a decimal fraction to make a comparison, there is always little bit more remaining. This time, however, the remaining amounts do not follow a repetitive pattern.

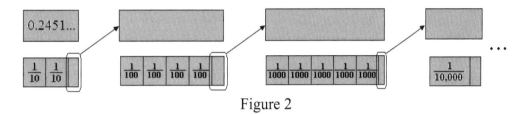

Figure 2

1. Suppose each of the following decimals represents the first three digits of an irrational number. Use fractions bars to draw a diagram like Figure 2 for each decimal.
   a. 0.213...   b. 0.101...   c. 0.432...

2. In general, it is difficult to know whether decimals like the ones in the last problem are actually irrational. However, some numbers can be shown to be irrational. For instance, consider the $\sqrt{2}$.
   a. Use a calculator to find a decimal representation for $\sqrt{2}$. What is it?
   b. What do you observe about this decimal representation of $\sqrt{2}$?
   c. Can we use a calculator to determine whether $\sqrt{2}$ is irrational? Why or why not?
   d. Read the proof in Section 8.1 of the textbook that shows $\sqrt{2}$ is irrational. Use your own words to explain why the proof works.

4. Another irrational number is π, which is found by taking the ratio of the circumference of any circle to its diameter. Search the Internet for a website that will calculate π up to 1,000,000 decimal places. First calculate π to 100 decimal places, then to 1,000 decimal places, then to 10,000 decimal places, and on so up to 1,000,000 decimal places.
   a. What do you observe about the decimal representation of π after each calculation?
   b. Does calculating π to 1,000,000 decimal places prove that π is irrational? Why or why not?

Part C: Summary and Connections

1. How are repeating decimals and irrational numbers similar? How are they different?

2. Students often have trouble understanding the relative sizes of large numbers. How could a computer-generated decimal representation of π to help students understand the relative size of 10,000 to 100,000? 100,000 to 1,000,000? Other powers of 10?

Exploration 8.4 Where is One For You?

Purpose: Students use a game involving base-ten blocks to develop algorithms for adding and subtracting decimals.

Materials: Base-ten blocks, a game score sheet, and spinners for the game

Part A: Adding Decimals

Find a partner to play the game, "Where is one for you?" To play the game you will need a set of base-ten blocks. You will also need Spinner A, Spinner B, and a game score sheet which are located on the page that follows this exploration. Each player begins the game with a flat, which is used to track the player's results after each turn. A flat has a value of 1, a rod has a value of 0.1, and a small cube has a value of 0.01.

1. Player 1 begins the game by spinning Spinners A and B and places the base-ten blocks equal to the sum of the spins on his flat. The spin and the total are then recorded on the score sheet. Player 2 then spins the spinners, places the base-ten equal to the sum of the spins on his flat, and records the spin and the total on the score sheet. Play continues with each player adding new blocks to his flat after each spin, making appropriate trades to keep a minimal collection of blocks on his flat. The player must explain any trades each time one is made. The first player to make a 1 by covering his flat wins. A player can spin both spinners or just one or the other on any given turn.

2. Modify the game so that the winner is the player who is closest to 1 without going over on a predetermined number of turns.

3. Modify the game so that the winner is the player who has the greatest value after a predetermined number of turns.

4. Describe the procedures you used to add decimals while playing these games. How does keeping track of the score with base-ten blocks help in adding the decimals?

Part B: Subtracting Decimals

The game can also be played with subtraction. Again, each player begins with a flat that has a value of 1.

1. Conduct play as in Part A, but this time the amount of the spin represents the blocks that are to be removed from the flat. Make exchanges as necessary so that blocks can be

removed. Play continues until a player is able to obtain a value of zero. After each turn, the amount of the spin must be recorded, exchanges explained, and the total verified.

2. Modify the game by letting the large cube have a value of 1, the flat have a value of 0.1, the rod have a value of 0.01, and the small cube have a value of 0.001. Use Spinners A, B, and C. On each turn a player can choose to spin Spinners A, B, and C, any two of the spinners, or any one spinner. A spinner cannot be used more than once on any single turn.

3. Describe the procedures you used to subtract decimals while playing these games. How does keeping track of the score with base-ten blocks help in subtracting the decimals?

Part C: Summary and Connections

1. What did you learn about addition and subtraction of decimals by playing "Where is one for you?"

2. Why does it make sense to "line up" the decimal points when adding or subtracting decimals? What mathematical concept is observed when we "line up" the decimal points when adding or subtracting decimals?

"Where is One for You?" Spinners

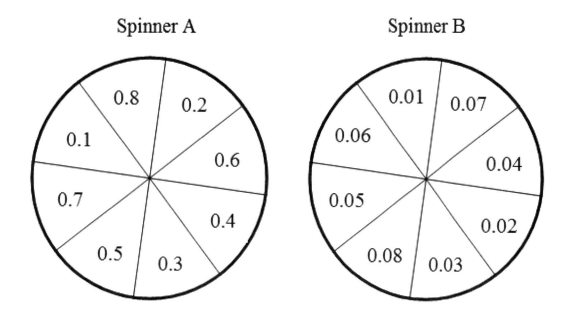

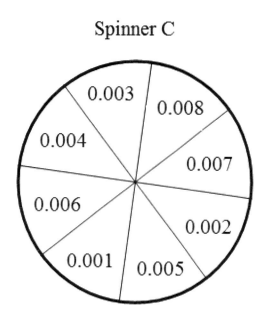

"Where is One For You?" Game Sheet

Player 1 Spin	Player 1 Total	Player 2 Spin	Player 2 Total

Exploration 8.5 Multiplying Decimals

Purpose: Students use models to develop an algorithm for multiplying decimals.

Materials: Base-ten blocks, decimal grid paper, and colored pencils

Part A: Multiplying Decimals with Base-Ten Blocks and the Set Model

We can use base-ten blocks and the set model to compute decimal products. Let a flat represent 1, a rod represent 0.1, and a small cube represent 0.01.

1. Consider the product $0.1 \cdot 0.3$. Like the product of two fractions, we can think of the product $0.1 \cdot 0.3$ as taking 0.1 of 0.3.
   a. Draw a diagram that shows how to use base-ten blocks to represent 1 group of 0.3.
   b. How can you show taking 0.1 of this set of base-ten blocks?
   c. What set of base-ten blocks is the product $0.1 \cdot 0.3$ equivalent to? How does this set compare to the flat and its value of 1?

2. Use a set of base-ten blocks to model each product. In each case, draw a diagram that shows how you used base-ten blocks to compute the product.
   a. $0.3 \cdot 0.2$ b. $1.3 \cdot 0.4$ c. $0.6 \cdot 3$ d. $0.2 \cdot 2.4$

3. Reconsider the products from Problem 2. In each case, compare the location of the decimal in the product to the location of the decimals in the factors. Use the patterns you see to suggest an algorithm for finding the product of two decimals.

Part B: Multiplying Decimals with Grid Paper

We can also use decimal grid paper to compute decimal products. Decimal grid paper consists of a grid of squares that are further subdivided into 100 smaller squares.

1. The following diagram shows how to represent the product $2 \cdot 0.3$ on decimal grid paper. Use what you see to answer each question.

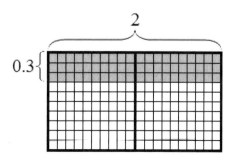

a. What area(s) in the diagram represent(s) a value of
   i. 1?   ii. 0.1?   iii. 0.01?
b. How is the product represented in the diagram?
c. What is the value of 2 · 0.3? How did you determine this value?

2. Use decimal grid paper to represent and compute each product.
   a. 0.3 · 0.2   b. 1.3 · 0.4   c. 0.6 · 3   d. 0.2 · 2.4

3. Reconsider the products from Problem 2. In each case, compare the location of the decimal in the product to the location of the decimals in the factors. Use the patterns you see to suggest an algorithm for finding the product of two decimals.

4. Decimal grid paper can also be used to compute products involving decimals with digits in the hundredths place values. What changes would have to be made to the values of the larger and smaller squares in order to use decimal grid paper to compute 0.3 · 0.07?

Part C: Summary and Connections

1. Use what you have learned in Parts A and B to write an algorithm for computing the product of two decimals.

2. Record your answers from Problem 2 of Part B in the first column of the following table. Use these answers to compute the products in Column 2.

Column 1	Column 2
0.3 · 0.2 =	3 · 2 =
1.3 · 0.4 =	13 · 4 =
0.6 · 3 =	6 · 3 =
0.2 · 2.4 =	2 · 24 =

a. How are the problems in Column 2 related to the problems in Column 1?
b. How could you use the products in Column 2 to find the products in Column 1?
c. What does this imply about using whole-number products to compute decimal products?

Exploration 8.6 Connecting Decimal Operations to Fraction Operations

Purpose: Students use fraction operations to understand the placement of the decimal point when adding, subtracting, and multiplying decimals.

Materials: Paper and pencil

Part A: Placing the Decimal When Adding or Subtracting Decimals

We can use fraction addition and subtraction to help us understand how to place the decimal point in sums and differences. We can do so in at least two ways.

1. One way to add or subtract decimals is to convert them to decimal fractions, find a common denominator, add or subtract, and then convert the result back to a decimal number. Compute each sum or difference in this way. Express your answer as a decimal.
    a. $0.3 + 0.6$ b. $5.2 - 0.9$ c. $4.31 + 2.2$ d. $7.1 - 8.03$

2. Compare your answers from Question 1. Write a sentence or two that summarizes how the position of the decimal in the sum or difference compares to the position of the decimal in the original two numbers.

3. Another way to add or subtract decimals is to write each number in expanded notation with decimal fractions, and then add or subtract the fractions with common denominators, making exchanges or regrouping as necessary. Compute each sum or difference in this way. Express your answer as a decimal.
    a. $4.31 + 2.5$ b. $3.48 - 2.36$ c. $15.945 + 6.701$ d. $23.074 - 16.305$

4. Compare your answers from Question 3. Write a sentence or two that summarizes how you placed the decimal point in the sum or difference using your results from adding or subtracting in expanded notation.

5. We can adapt the standard algorithms for whole-number addition and subtraction to add and subtract decimals. Specifically, the algorithms for decimals begin by lining up the decimal points in the numbers. The sum or difference is computed using the whole-number algorithms, and then the decimal point is brought straight down to place it in the answer. Use what you have learned from the last several questions to explain why the decimal point can be placed in this way.

Part B: Placing the Decimal When Multiplying Decimals

One way to multiply decimals is to convert each decimal to a decimal fraction and then multiply using fraction multiplication.

1. For each problem, write the decimals as decimal fractions and then multiply the fractions. Convert the product back to a decimal.
    a. $0.4 \cdot 0.7$ b. $2.5 \cdot 0.7$ c. $4.8 \cdot 1.7$ d. $9.33 \cdot 4.6$

2. Compare your answers from Question 1. Write a sentence or two that summarizes how the position of the decimal in the product is related to the position of the decimals in the original two numbers.

3. We can adapt the standard algorithm for whole-number multiplication to multiply decimals. Specifically, the algorithm for decimals begins by multiplying the decimals as if they were whole numbers. The decimal point is then placed in the product so that the product has the same number of decimal places as the total number of decimal places in the factors. Use what you have learned in the last two questions to explain why this algorithm for multiplying decimals makes sense.

Part C: Summary and Connections

1. Is it possible to use fraction division to explain the placement of the decimal point in decimal division? Explore this question with several example quotients. Write a short paragraph that summarizes what you find.

2. a. Whenever we compute with decimals, we almost always use terminating decimals. Why do you think this is the case?
   b. Is it possible to adapt the standard algorithms for the whole-number operations to add, subtract, and multiply repeating decimals? Explore this question with several example problems. Write a short paragraph that summarizes what you find.

3. Do you think it would make sense to use fraction operations to explain decimal operations in the elementary or middle grades classroom? Why or why not?

Exploration 8.7 Student Errors in Using Scientific Notation

Purpose: Students examine common errors that children make when using scientific notation.

Materials: Paper and pencil

Part A: Student Errors with Understanding Scientific Notation

As children learn scientific notation, they often make mistakes not just in writing the notation, but also in understanding the relative sizes of numbers the notation represents. Examine each situation and identify the mistake the child has made. Explain how you might correct the student's thinking.

1. Thomas is writing numbers in scientific notation. He writes 7,500 as 7.5^3 and 84,000 as 8.4^4.

2. Imam is writing numbers in scientific notation. He writes 4,560 as 456×10 and 73,000 as 73×10^3.

3. Alli is writing numbers in scientific notation. She writes 0.0056 as 5.6×10^3 and 0.000944 as 9.44×10^4.

4. Akeem is converting numbers from scientific notation to decimal notation. He writes 7.11×10^3 as 0.00711 and 6.89×10^{-4} as 68,900.

5. Jessie is ordering numbers written in scientific notation. She claims that 2.41×10^6 is less than 5.69×10^3 because $2.41 < 5.69$.

6. Taylor states that 9.1×10^4 is about three times as big as 3.05×10^2.

Part B: Student Errors with Operations in Scientific Notation

Children also make mistakes when computing with numbers written in scientific notation. Examine the problem solved by each student. Identify the mistake and explain how you might correct the student's thinking.

1. Dallas: $(4.1 \times 10^3) + (3.65 \times 10^3) = 7.75 \times 10^6$

2. Morgan: $(8.6 \times 10^5) - (3.7 \times 10^3) = 4.9 \times 10^2$

3. Sanjay: $(5.8 \times 10^7) + (8.33 \times 10^2) = 14.13 \times 10^2$

4. Heather: $(1.8 \times 10^3) \times (4.09 \times 10^3) = 7.362 \times 10^3$

5. Nikki: $(3.5 \times 10^4) \times (2.1 \times 10^5) = 7.35 \times 10^{20}$

6. Hudson: $(6.11 \times 10^3) \times (7.8 \times 10^6) = 4.7658 \times 10^9$

7. Isaiah: $(8.4 \times 10^6) \div (2.1 \times 10^2) = 4 \times 10^2$

Part C: Summary and Connections

1. Why do you think many children find scientific notation difficult to master?

2. Children often use calculators to help them make computations with scientific notation.
    a. What are some common errors that children might make when using a calculator to make computations with numbers written in scientific notation?
    b. How might you, as a teacher, help children to avoid these mistakes?

Exploration 8.8 Understanding Proportional Reasoning Problems

Purpose: Students examine and solve two types of proportional reasoning problems.

Materials: Paper and pencil

Part A: Missing-Value Problems

Proportional reasoning problems generally fall into two categories: *missing-value* problems and *comparison* problems. In missing-value problems, three of the four numbers in the proportion are given and the task is to find the missing number.

1. Hannah is 56 in. tall and Ashley is 52 in. tall. When measured in meters, Hannah is 1.42 m tall. How tall is Ashley in meters?
    a. Why is this a proportional reasoning problem? Why is it a missing-value problem?
    b. Use a table to represent the proportion.
        i. What are different ways to make the table and still express the same proportional relationship?
        ii. Why are the labels important in setting up the table correctly?
        iii. How is a variable useful in expressing the proportional relationship?
    c. Use your table to solve the problem.

2. Cole can run 3 miles in 21 minutes. How long will it take him to run 10 miles at the same speed?
    a. Why is this a proportional reasoning problem? Why is it a missing-value problem?
    b. Use a unit rate to solve the proportion. Why is a unit rate an effective way to solve the problem?

3. The ratio of male students to female students at a particular university is 4 to 5. If 3,580 men are enrolled in the university, how many women are enrolled?
    a. Why is this a proportional reasoning problem? Why is it a missing-value problem?
    b. Use an equation to represent the proportion.
        i. What are different ways to write the equation and still express the same proportional relationship?
        ii. Why are the labels important in setting up the equation correctly?
        iii. The equation can be solved by cross-multiplying. What happens to the units in the cross product?
    c. Use your equation to solve the problem.

Part B: Comparison Problems

Another type of proportional reasoning problem is a comparison problem. In this case, two sets of values are given and the task is to determine whether the ratios are equal or whether one is larger than the other.

1. One pitcher of lemonade was made with 10 cups of water and 12 lemons. Another pitcher was made with 15 cups of water and 18 lemons. Does one pitcher of lemonade have a stronger lemon taste than the other or are they the same?
    a. Why is this a proportional reasoning problem? Why is it a comparison problem?
    b. Use a picture to solve the problem. Why is a picture an effective way to solve the problem?

2. Timothy can drive his car 330 miles on 15 gallons of gas. Tamara can drive 276 miles on 12 gallons of gas. Which car is more fuel efficient?
    a. Why is this a proportional reasoning problem? Why is it a comparison problem?
    b. Use a unit rate to solve the problem. Why is this an effective way to solve the problem?

3. Dr. Cohen has two mathematics classes for elementary teachers. In one class, 19 of 22 students passed. In the other, 21 of 25 passed. Which class had a better passing rate?
    a. Why is this a proportional reasoning problem? Why is it a comparison problem?
    b. Compare two fractions to solve the problem. Why is this an effective way to solve the problem?

Part C: Summary and Connections

1. a. Explain in your own words how missing-value problems are different from comparison problems.
   b. If you were given a proportional reasoning problem that you had never solved before, how would you know how to classify it as a missing-value or as a comparison problem?

2. a. Which strategies do you think are the easiest to use when solving proportion problems?
   b. What other strategies have you used or could you use to solve
      i. missing-value problems?   ii. comparison problems?
   c. Which strategies do you think are the easiest to use? How might this influence the way you teach children to solve proportional reasoning problems?

Exploration 8.9 Solving Proportion Problems with the Bar Model

Purpose: Students use the bar model to solve proportional reasoning problems.

Materials: Paper and pencil

Part A: Using the Bar Model to Solve Proportion Problems

A bar model can be used as a visual representation of the relationships given in a proportion problem. For instance, consider the following problem:

> At lunch, 360 students ate in the cafeteria. They could choose either a hamburger or a hot dog. The ratio of students who chose a hamburger to a hot dog was 5 to 3. How many students chose a hamburger? How many chose a hot dog?

To model the problem, we use a single rectangle to represent a group of students. Even though we do not know how many students are in a group, the ratio tells us that 5 groups of students got hamburgers and 3 groups of students got hot dogs. Because the total number of students who chose either a hamburger or a hot dog is 360, we can make the following diagram:

Students who chose a hamburger [5 boxes]
Students who chose a hot dog [3 boxes]
} 360 students

From the diagram, we can see that the 8 groups represent 360 students, so each group must represent 360 ÷ 8 = 45 students. Consequently, the number of students who ordered a hamburger is 5 groups of 45, or 5 · 45 = 225 students; the number of students who ordered a hot dog is 3 groups of 45, or 3 · 45 = 135 students. Checking our work, we have 225 + 135 = 360 students and the ratio of 225 to 135 is equivalent to a ratio of 5 to 3.

Use the bar model to solve each problem. In each case, draw a diagram of your model and be sure to identify what each component of the model represents.

1. Henri cut a rope into two pieces with lengths having a ratio of 5 to 2. The shorter piece is 70 cm long. What is the length of the original rope?

2. The ratio of the cost of a shirt to the cost of a pair of pants is 4 to 6. If the pair of pants costs $15 more than the shirt, find the cost of each.

3. The ratio of the number of girls to the number of boys at an event is 2 to 7. If there are 140 boys, how many girls are attending?

4. Scott is making iced tea by mixing tea concentrate and water in the ratio of 2 to 7. If he uses 4 pints of concentrate, how much water should he use?

5. Claire and Eve shared $520 in the ratio of 8 to 5. How much more money did Claire receive than Eve?

6. The blue paint that I have is too dark. I have some white paint that I can use to lighten it. I am mixing paint in the ratio of 4 parts blue to 1 part white. If it takes 3 gallons to paint a room, how many gallons of each color should I mix together so that no mixed paint is left over?

Part B: Summary and Connections

1. The bar model can also be used to help us represent proportion problems symbolically. For instance, in the example problem in Part A, we can let x represent a group of students rather than a rectangle. Then $x + x + x + x + x = 5x$ represents the number of students who chose a hamburger and $x + x + x = 3x$ represents the number of students who chose a hot dog. Because the total number of students is 360, then $5x + 3x = 360$.
    a. What is the value of x?
    b. How would you find the number of students who chose a hamburger? A hot dog?
    c. What is the connection between the symbolic representation and the bar model representation?

2. Use the bar model to help you solve each of the problems in Part A symbolically.

Exploration 8.10 The Golden Ratio and the Human Body

Purpose: Students use measurement to discover how the Golden Ratio occurs in the human body.

Materials: Yardsticks, tape measures, string, or other measuring tools

Part A: Finding the Golden Ratio in the Human Body

Form a small group with four or five of your peers. Use measuring tools to measure the following lengths for every person in the group. Record your data in the table.

Measurement 1: Height without shoes.
Measurement 2: Top of head to fingertips when arms are resting at sides.
Measurement 3: Top of the head to navel.
Measurement 4: Width of shoulders.
Measurement 5: Length of head.

Name	Measure 1	Measure 2	Measure 3	Measure 4	Measure 5

1. For each person in the group, compute the ratio of the height to the distance between the top of the head and the fingertips. That is, compute the ratio of Measurement 1 to Measurement 2. Record the information in the following table. Does the ratio of these two measures tend towards a particular value? If so, what is it?

Name					
$\dfrac{\text{Measurement 1}}{\text{Measurement 2}}$					

2. For each person in the group, compute the ratio of the distance between the head and the fingertips to the distance between the head and the navel. That is, compute the ratio of Measurement 2 to Measurement 3. Record the information in the following table. Does the ratio of these measurements tend towards a particular value? If so, what is it?

Name					
Measurement 2 / Measurement 3					

3. For each person in the group, compute the ratio of the distance between the head and the navel to the width of the shoulders. That is, compute the ratio of Measurement 3 to Measurement 4. Record the information in the following table. Does the ratio of these measurements tend towards a particular value? If so, what is it?

Name					
Measurement 3 / Measurement 4					

4. For each person in the group, compute the ratio of the width of the shoulders to the length of the head. That is, compute the ratio of Measurement 4 to Measurement 5. Record the information in the following table. Does the ratio of these two measurements tend towards a particular value? If so, what is it?

Name					
Measurement 4 / Measurement 5					

5. Compare the results for each set of ratios you have computed. What do you notice? Is the result surprising?

6. Continue to explore by considering other ratios using these five measurements. Write a short summary of what you find.

Part B: Summary and Connections

1. The values you obtained in the first four questions of Part A should all be approximately 1.62. This number is called the Golden Ratio. It is an irrational number that occurs in a surprising number of places. Search the Internet to discover the background of this interesting number.

2. Search the Internet for other ways that the Golden Ratio occurs in the body. Write a short report on what you find.

Exploration 8.11 Student Errors in Proportional Reasoning

Purpose: Students examine common errors that children make when solving proportional reasoning problems.

Materials: Paper and pencil

Part A: Student Errors in Proportional Reasoning

Suppose Mrs. Hunter, a sixth grade teacher, gives her students the following problem:

Jalinda purchases 3 pieces of candy for 25¢. How much will it cost her to purchase 12 pieces of candy?

1. a. Use a unit rate to solve the problem.
   b. Use cross-multiplication to solve the problem.
   c. What other strategies can you think of to solve the problem?

2. Sasha claims that because 3 pieces cost 25¢, then 12 pieces must cost 12 · 25 = $3.00. What did Sasha do wrong and how might you correct her thinking?

3. When Jacori solves the problem, he notices that 25 is 22 more than 3. So he concludes that 12 pieces must cost 12¢ + 22¢ = 34¢. What did Jacori do wrong and how might you correct his thinking?

4. China uses the proportion $\frac{3 \text{ pieces}}{25¢} = \frac{x¢}{12 \text{ pieces}}$ and claims that 12 pieces will cost 6.25¢. What did China do wrong and how might you correct her thinking?

5. Timothy says that if 3 pieces cost 25¢, then one piece will cost 12¢. Consequently, 12 pieces will cost 12 · 0.12 = 1.44¢? What did Timothy do wrong and how might you correct his thinking?

Part B: Summary and Connections

1. a. What other kinds of errors might children make when solving proportional reasoning problems?
   b. Why do you think so many children have trouble solving these kinds of problems?

2. Give some examples of how we use proportional reasoning in our daily lives. What does this suggest about the importance of teaching this topic?

Exploration 8.12 Solving Percent Problems with the Bar Model

Purpose: Students use the bar model to solve percent problems.

Materials: Paper and pencil

Part A: Representing Percents with a Bar

One way to solve percent problems is to use the bar model. To do so, we let the length of the bar represent the whole, or 100%, and then shade the appropriate amount to represent the needed part. If we are given a shaded region and asked to find the percent, we can think of the region as a fraction of the whole and then convert it to a percent.

For each of the following, write the part of the bar that is shaded as a fraction in simplest form and as a percent.

1.

Fraction: _____ Percent: _____

2.

Fraction: _____ Percent: _____

3.

Fraction: _____ Percent: _____

4.

Fraction: _____ Percent: _____

Part B: Using the Bar Model to Solve Percent Problems

The bar model can be used to represent the relationships given in a percent problem. For instance, consider the following problem in which the percent and the whole are known:

Thirty percent of a company's 180 employees are women. How many women does the company employ?

To model the problem, we use a single bar to represent all of the 180, or 100%, of the company's employees. Because 30% of the employees are women, we subdivide the bar into ten equivalent pieces and shade three of them. Each piece is 10% of the total, so it must represent 18 employees. Consequently, 30% of the bar is 3 groups of 18 employees, or 54 female employees.

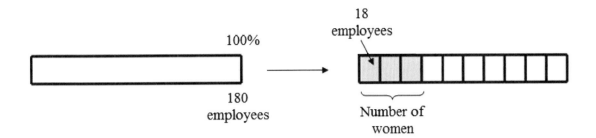

Use the bar model to represent and solve each percent problem. In each case, draw a diagram of your model and be sure to identify what each component of the model represents.

1. In the Powdersville school district, 36 schools offer after-school programs for their students. These 36 schools represent 75% of the schools in the county. How many schools are in the Powdersville school district?

2. A jacket is on sale for 30% off the listed price of $89.50. What is the new price of the jacket?

3. A small business made a profit of $27,800 during its first year of operation. In the second year, its profit increased by 60%. What was the profit for the second year?

4. Mr. Smith agreed to donate 40 acres of land to the city provided they use 8 acres for a park. What percent of the donated land will be used for the park?

5. The selling price of a chair is $430. This price represents a mark-up of 25% of the wholesale price. Find the wholesale price.

Part C: Summary and Connections

1. Most percent problems generally involve three values: the percent, the part, and the whole. How are each of these numbers represented in the problems above?

2. What ideas must students understand about the concept of a percent to use the bar model effectively?

3. What percents do not work well with the bar model? Why do they not work well?

Exploration 8.13 Grading with Points and Percents

Purpose: Students explore the differences between using points and percentages when determining classroom grades.

Materials: Calculator, paper, and pencil

Part A: Points Versus Percentages

Teachers in the elementary grades often use two schemes to determine grades. In both schemes, graded assignments are given a certain number of points and those points are turned into percentages. In one scheme, the points for each assignment are turned into a percentage, and the percentages are averaged to determine a student's final grade. Because each grade is given as a percentage, every grade has the same weight in the final grade. In the other scheme, the points from each assignment are added and then divided by the total number of points to find the percentage for a student's final grade. Because each assignment has a different number of points relative to the total, each assignment has a different weight in the student's final grade. As we will see, these grading schemes can lead to different final grades.

Consider the following problems in which the grades earned by three fourth grade students are given. The students are in the same class and have completed the same six homework assignments.

1. Alissa earns the following points on the six assignments. Calculate the percentage for each assignment, the total number of points earned by Alissa, and the total number of points given on the assignments.

	H1	H2	H2	H4	H5	H6	Total Points
Points Earned	9	41	37	38	32	13	
Points on Assignment	10	45	40	40	35	15	
Percentage							

   a. Compute Alissa's final homework grade by averaging the percentages for the six assignments.
   b. Compute Alissa's final homework grade by taking the total points she earned and dividing it by the total points for all six assignments.
   c. How do the answers for the two methods compare?

2. Isaiah earns the following points on the six assignments. Calculate the percentage for each assignment, the total number of points earned by Isaiah, and the total number of points given on the assignments.

	H1	H2	H2	H4	H5	H6	Total Points
Points Earned	3	43	38	36	31	4	
Points on Assignment	10	45	40	40	35	15	
Percentage							

    a. Compute Isaiah's final homework grade by averaging the percentages for the six assignments.
    b. Compute Isaiah's final homework grade by taking the total points he earned and dividing it by the total points for all six assignments.
    c. How do the answers for the two methods compare? What is the cause of the discrepancy between the two grades?
    d. Which grade do you think best represents Isaiah's performance in class?

3. Morgan earns the following points on the six assignments. Calculate the percentage for each assignment, the total number of points earned by Morgan, and the total number of points given on the assignments. Use the table to answer the questions.

	H1	H2	H2	H4	H5	H6	Total Points
Points Earned	10	30	25	28	21	15	
Points on Assignment	10	45	40	40	35	15	
Percentage							

    a. Compute Morgan's final homework grade by averaging the percentages for the six assignments.
    b. Compute Morgan's final homework grade by taking the total points she earned and dividing it by the total points for all six assignments.
    c. How do the answers for the two methods compare? What is the cause of the discrepancy between the two grades?
    d. Which grade do you think best represents Morgan's performance in class?

Part B: Summary and Connections

1. In this scenario, there were more grades worth a large number of points than there were grades worth a small number of points. Explore other grading situations involving assignments worth a different number of points. Are there discrepancies between the two grading systems in the situations you explore? Explain.

2. Of the two methods, which do you think is the best way to determine a final grade? Why?

Chapter 9
Algebraic Thinking

It may surprise you that **algebraic thinking** is advocated at all grade levels. However, there is a distinct difference between algebraic thinking and the formal algebra courses taught in high school. In the elementary and middle grades, algebraic thinking includes three areas: functional thinking, general arithmetic, and mathematical modeling.

Functional thinking is the process we use to identify, represent, and extend the patterns and relationships that exist between sets of objects. In school, children develop functional thinking as they describe qualitative and quantitative changes in simple patterns that involve colors, shapes, sounds, and even daily routines. Once they are adept with these simple patterns, children begin to use different representations to work with patterns of change between sets of numbers.

General arithmetic encompasses two areas. One involves identifying and extending regular patterns embedded in the numerical operations. The other uses the properties of arithmetic to find unknown quantities. Children first work with this aspect of algebraic thinking as they learn to compute whole-number operations. They continue to use it throughout the middle grades as they learn to solve simple equations.

Mathematical modeling is the process of creating and using a mathematical representation to solve a real-world problem. Mathematical models come in a variety of forms and can use both functional thinking and general arithmetic. Children first make mathematical models when they use what they know about arithmetic to solve simple story problems. As they progress, students continue to learn and use new representations to solve problems.

The explorations in Chapter 9 use a variety of representations to work through the different aspects of algebraic thinking. Specifically,

- Explorations 9.1 - 9.3 use numerical sequences to develop the notion of functional thinking.
- Explorations 9.4 - 9.9 continue functional thinking by working with ways of representing functions and linear functions.
- Explorations 9.10 - 9.11 focus on solving equations and the common mistakes students make while doing so.
- Explorations 9.12 - 9.13 use mathematical modeling to solve problems.

Exploration 9.1 Understanding Arithmetic Sequences

Purpose: Students use contextual problems to understand arithmetic sequences.

Materials: Paper and pencil

Part A: An Increasing Arithmetic Sequence

An **arithmetic sequence** is a list of numbers in which each successive term is found by adding or subtracting a fixed number to the previous term. Many arithmetic sequences are increasing; that is, the numbers in the sequence continue to grow larger. To better understand arithmetic sequences, consider the following problem and subsequent questions.

Marcie recently opened a savings account with an initial deposit of $1,500. She plans to save $250 a month. She sets up the following table to track how much she has saved in the account.

Month	1	2	3	4	5	6	7
Amount in Account	$1,500	$1,750					

1. Complete the table by filling in the amount of money in the account for each month.

2. Each **term**, or number, in the sequence has both a position and a value. That is, a term is first, second, third, or so on.
    a. What part of the table designates the positions of the terms in the sequence?
    b. What part of the table designates the values of the terms in the sequence?

3. a. How does the value of the account increase from month to month?
   b. The sequence increases by the same value each month. This value is called the **common difference**. Use the common difference to find the amount in the account for months 8 through 12?
   c. In general, how can we find the common difference of any arithmetic sequence?
   d. What would happen to the sequence if we changed the common difference to $350? What about $150? What does this imply about the role of the common difference in defining an arithmetic sequence?

4. The first term in any arithmetic sequence is also important.
    a. What is the first term of the sequence?
    b. What would happen to the sequence if Marcie had initially deposited $2,000? Initially deposited $1,000? What does this imply about the role of the first term in defining the sequence?

Part B: A Decreasing Arithmetic Sequence

Arithmetic sequences can also be decreasing. That is, the terms of the sequence get smaller by the same constant amount. Consider the following problem.

Tamiru has $100 to purchase music from an on-line music provider. Each song costs $1.29 and he plans to buy two songs every week. He sets up the following table to keep track of how much of the $100 he has spent.

Week	1	2	3	4	5	6	7
Amount in Account	$100	$97.42					

1. Complete the table by filling in the amount in the account for each week. Is the sequence of the remaining dollar amounts an arithmetic sequence? How do you know?

2. What is the first term of the sequence?

3. What is the common difference of the sequence? Why does the common difference make the sequence a decreasing sequence?

4. Use the common difference to find the next five terms of the sequence.

Part C: The General Term of an Arithmetic Sequence

The last two problems show that arithmetic sequences always follow a pattern of constant change. This makes it is possible to write a general term for the sequence that is represented with an algebraic expression that connects the position of a term, n, to its value, a.

1. To find the expression for the general term, we can examine the terms of a sequence for a pattern. Use the sequence from Part A to complete the following table.

Term	Position	Values related to the first term and common difference
First	1	1,500
Second	2	1,750 = 1,500 + 250 = 1,500 + 1(250)
Third	3	2,000 = 1,500 + 250 + 250 = 1,500 + 2(250)
Fourth	4	
Fifth	5	
General	n	

2. a. What algebraic expression do you get for the general term?
   b. The general term is useful because it connects the position of a term to its value. If we know the position of a particular term, we can substitute it into the general term to find its value. Use the general term of the sequence to find the amount in the account after 10 months, 15 months, and 2 years.
   c. How many months will it take for the account to have $5,000?

3. a. Find the general term for the arithmetic sequence in Part B. Use it to find the amount of money in the account after the 15th, 20th, and 25th weeks.
   b. How many weeks will it take before Tamiru no longer has enough money to purchase music?

Part D: Summary and Connections

1. a. What are the two defining characteristics of an arithmetic sequence?
   b. What determines whether an arithmetic sequence is increasing or decreasing?

2. Make a list of three or four arithmetic sequences that you have encountered in your everyday life. Did you use the general term for these sequences? If so, how?

Exploration 9.2 Understanding Geometric Sequences

Purpose: Students use contextual problems to understand geometric sequences.

Materials: Spreadsheet, paper, and pencil

Part A: An Increasing Geometric Sequence

A **geometric sequence** is a list of numbers in which each successive term is found by multiplying or dividing a fixed number to the previous term. To better understand geometric sequences, consider the following problem and subsequent questions.

Trevor created a website for a new running club in an urban area. It only had 200 hits in its first month, but he expects the number of hits to double every month. Trevor sets up the following table to track the number of hits to the site each month.

Month	1	2	3	4	5	6	7
Number of Hits	200	400					

1. Assume Trevor is correct about how the number of hits on the site will increase. Complete the table by filling in the number of hits for each month.
    a. What part of the table designates the positions of the terms in the sequence?
    b. What part of the table designates the values of the terms in the sequence?
    c. Why does the sequence for the number of hits represent a geometric sequence?

2. The sequence increases by the same factor each month. This value is called the **common ratio**.
    a. Use the common ratio to find the number of hits for months 8 through 12.
    b. In general, how can we find the common ratio of any geometric sequence?
    c. What would happen to the sequence if we changed the common ratio to 3? What about $\frac{1}{2}$? What does this imply about the role of the common ratio in defining the sequence?

3. The first term in any geometric sequence is also important.
    a. What is the first term of the sequence?
    b. What would happen to the sequence if the site initially had 1,000 hits? What about 50 hits? What does this imply about the role of the first term in defining the sequence?

Part B: The General Term of a Geometric Sequence

The problem in Part A shows that the terms in a geometric sequence always change by the same factor. This makes it possible to use an algebraic expression to write a general term for the sequence that connects the position of a term, n, to its value, a.

1. To find the expression for the general term, we can examine the terms of a sequence for a pattern. Use the sequence from Part A to complete the following table.

Term	Position	Value in terms of first term and common difference
First	1	200
Second	2	$400 = 200 \cdot 2$
Third	3	$800 = 200 \cdot 2 \cdot 2 = 200 \cdot 2^2$
Fourth	4	
Fifth	5	
General	n	

2. a. What algebraic expression do you get for the general term?
   b. The general term is useful because it connects the position of a term to its value. If we know the position of a particular term, we can substitute it into the general term to find its value. Use the general term to find the number of hits to the website after 10 months, 15 months, and 2 years.
   c. How long will it take for the website to have over 1,000,000 hits?

Part C: Using a Spreadsheet

In the last example, we used the general term to find the values of only a few terms. In some situations, however, we need to generate the values of many terms. We can use a computer program called a **spreadsheet** to make the computations both quickly and efficiently. To do so, we first enter the positions of the terms in column A of the spreadsheet. We can do this directly or with the FILL DOWN command. To use the FILL DOWN command, enter 1 in cell A1. Then enter the formula "= A1 + 1" into cell A2. This tells the spreadsheet to add 1 to the value in cell A1 and place the result in cell A2. From cell A2, fill down until the needed number of cells is reached. The spreadsheet will automatically perform any calculations.

In column B, we can calculate the values of the terms by entering the formula for the general term in cell B1. In this case, we replace the variable n with A1, which tells the spreadsheet to get the position of the term from cell A1. Again, fill down and the spreadsheet will automatically place the value of the term next to its position.

Use a spreadsheet to solve the following problems.

1. A ball is dropped form the top of a building that is 81 ft tall. Each time it bounces, it only goes up to $\frac{1}{3}$ of the previous height.
    a. Why does the height of the ball after each bounce represent a geometric sequence? Write an expression for the general term of the sequence.
    b. Use a spreadsheet and the general term to find the height of the ball after the fifth bounce.
    c. How many times will the ball bounce before the height it reaches is less than 6 in.?
    d. If the ball stops bouncing after the tenth bounce, what is the total distance it travels before it stops?

2. Jenna purchased a car for $22,929. Each year, the value of the car depreciates by 15%.
    a. Why do the yearly decreases in the value of the car represent a geometric sequence? Write an expression for the general term of the sequence.
    b. Use a spreadsheet and the general term to find the value of the car after 5 years?
    c. How many years will it take for the value of the car to be less than $5,000?
    d. Will the value of the car ever reach zero? Explain.

Part D: Summary and Connections

1. a. What are the two defining characteristics of a geometric sequence?
   b. What determines whether a geometric sequence is increasing or decreasing?

2. Review the section in Chapter 8 that discusses interest. How is interest connected to geometric sequences?

3. Make a list of three or four other geometric sequences you have encountered in your everyday life. Did you use the general term for these sequences? If so, how?

Exploration 9.3 The Fibonacci Sequence and the Golden Ratio

Purpose: Students use technology to explore the relationship between the Fibonacci sequence and the Golden Ratio.

Materials: Calculator, spreadsheet, and Internet access

Part A: The Fibonacci Sequence

Take a few moments and read about the Fibonacci sequence in Section 9.1 of the textbook. Use what you read to answer the following questions.

1. What are the first ten terms of the Fibonacci sequence?

2. A sequence is recursively defined if the value of each term, except possibly for the first few terms, is based on the values of previous terms.
    a. Why is the Fibonacci sequence a recursively defined sequence?
    b. What is the general term of the Fibonacci sequence?
    c. Use the general term to generate the next ten terms of the Fibonacci sequence.

Part B: The Fibonacci Sequence and the Golden Ratio

The Fibonacci sequence has an interesting connection to the Golden Ratio. Consider the next several questions.

1. Consider the first eight numbers in the Fibonacci sequence. Use a calculator to complete the following table by computing the ratios of consecutive Fibonacci numbers. Express your answers in decimal form. What do you observe?

$\dfrac{F_n}{F_{n-1}}$	$\dfrac{F_2}{F_1}$	$\dfrac{F_3}{F_2}$	$\dfrac{F_4}{F_3}$	$\dfrac{F_5}{F_4}$	$\dfrac{F_6}{F_5}$	$\dfrac{F_7}{F_6}$	$\dfrac{F_8}{F_7}$
Decimal Value							

The last problem shows that the ratios of consecutive Fibonacci numbers tend towards a constant. We can investigate this further by using a spreadsheet. First, use the spreadsheet to generate the first 50 counting numbers. Begin by labeling the first column *Term Number*. Then in cell A2, enter "1." In cell A3, enter "=A2 + 1" and click on the check above the spreadsheet to enter the formula. Use the Fill Down feature to generate the first 50 counting numbers in the first column. Next, label the second column *Fibonacci Number* and enter "1" in both cell B2 and B3. In cell B4, enter "= B2 + B3" which tells the computer to add the numbers in B2 and B3. Click on the check and fill down to cell B51 to create the first 50 terms of the Fibonacci sequence. Finally, label the third column *Ratio*. In cell C2, enter "= B3/B2". Click on the check and fill down to cell C51. The computer will automatically compute the ratio of consecutive Fibonacci numbers, $\frac{F_{n+1}}{F_n}$, for values of *n* from 1 to 49.

2. a. As the terms of the sequence increase, what value does the ratio $\frac{F_{n+1}}{F_n}$ approach?

   b. Why does it make sense that the ratio would approach a constant?

3. Search the Internet for the decimal value found in Question 2. What is the number called and why is it special?

4. Return to the spreadsheet. Change the value in cell B2 to 4 and the value in cell B3 to 7.
   a. What do you observe about the values of the other cells?
   b. Now change cell B2 to 11 and cell B3 to 18. What happens to the other cells?
   c. Try other numbers in cells B2 and B3. Write down any patterns that you observe.

5. Return to the spreadsheet and change cell B2 and cell B3 back to their original value of 1. Go to cell C2 and change the ratio to $\frac{B4}{B2}$. Fill down in column C.
   a. What do you observe about the values of the other cells?
   b. How does the new ratio compare to the Golden Ratio?
   c. Use the spreadsheet to explore other ratios of the Fibonacci numbers. Write down any patterns that you observe.

Part C: The Fibonacci Sequence in Nature

You may find it surprising that the Fibonacci sequence and Golden Ratio appear in a number of places in nature.

1. Search the Internet to discover how the Fibonacci sequence appears in each of the following areas. In each case, write a short paragraph summarizing what you find.
   a. Flower petals   b. Pineapples and pine cones.   c. Sea shells.

2. Search the Internet to discover how the Golden Ratio appears in each of the following areas. In each case, write a short paragraph summarizing what you find.
   a. The human face.   b. The human body.   c. Butterflies and moths

Part D: Summary and Connections

1. Use the Internet to find other facts and properties about the Fibonacci sequence. Write a short report on what you find.

2. Use the Internet to search how the Golden Ratio is found in geometrical shapes. Write a short report on what you find.

Exploration 9.4 Understanding the Meaning of a Function

Purpose: Students use a real-world context to identify and understand functions.

Materials: Paper and pencil

Part A: Functions in a Context

A **function** from a set A to a set B is a rule of correspondence between two sets such that each element of the first set is paired with exactly one element of the second set. To better understand this definition, consider the following problems. In each case, let A be the set of United States citizens enrolled at your university this semester.

1. Let B be the set of days in the year, or $B = \{$Jan. 1, Jan. 2, Jan. 3, ... Dec. 31$\}$. Consider the relationship between each member of set A and his or her birthday. Is the relationship from A to B a function? Explain how you know.

2. Let set S be the set of 9-digit social security numbers. Consider the relationship between each member of set A and his or her social security number. Is the relationship from A to S a function? Explain how you know.

3. Let set P be the set of 10-digit phone numbers. Consider the relationship between each member of set A and his or her phone number. Is the relationship from A to P a function? Explain how you know.

4. Let set N be the set of 4-digit whole numbers. Consider the relationship between each member of set A and the last 4 digits of his of her social security number. Is the relationship from A to N a function? Explain how you know.

5. Let set D be the set of days of the week, or $D = \{$Sunday, Monday, ..., Saturday$\}$. Consider the relationship between each member of set A and days of the week on which he or she has class. Is the relationship from A to D a function? Explain how you know.

Part B: Domain, Range, and One-to-one Functions

The set A is called the **domain** of the function and it represents the set of all possible objects for which the rule of the function is meaningful. The **range** of the function is the subset of B that contains all objects in B that have been paired with something from the domain.

1. State the domain and range for each relationship in Part A that is a function.

2. A function is **one-to-one** if each element of the second set is paired with exactly one element from the first set. Which of the relationships in Part A represents a one-to-one function? Explain how you know.

Part C: Summary and Connections

1. List five other examples of real-world functions other than the examples given in Part A. In each case, state the domain and range.

2. If a relationship from set A to set B is a function, then we say each element in set A *uniquely determines* an element in set B. Explain what is meant by *uniquely determines*.

3. Refer to the definition of a function. Is it possible for set A to contain an element that is not paired to an element in set B? Why or why not?

Exploration 9.5 Functional Thinking and Patterns of Change

Purpose: Students learn to recognize functions as patterns of change by examining different representations of functions.

Materials: Paper and pencil

Part A: Functional Thinking and Patterns of Change

Functional thinking is the process we use to identify, represent, and extend patterns and relationships that exist *between* sets of objects. If the sets have numbers, then we can look for a pattern that indicates how the change in one set relates to or produces a change in the other. The rule that governs this pattern of change is called a **function**.

1. A function can be represented in a table in which the first column gives the domain values and the second column gives the range values. The rule of the function is indicated by particular pairings between the columns. Describe the function in each table. Then state the domain and range values used in the function.

   a. Function F

x	y
1	-4
2	-3
3	-2
4	-1
5	0

   b. Function G

x	y
1	-1
2	-4
3	-9
4	-16
5	-25

   c. Function H

x	y
0	0
1	-1
4	-2
9	-3
16	-4

2. A function can also be represented as a set of ordered pairs, in which the first number represents a domain value and the second a range value. The rule of the function is indicated by the particular pairings in the ordered pairs. Describe the function in each set of ordered pairs. Then state the domain and range values used in the function.
    a. Function J: {(1, 3), (2, 3), (3, 3), (4, 3), (5, 3)}
    b. Function K: {(0, 1), (1, 2), (2, 4), (3, 8), (4, 16)}
    c. Function L: {(0, 0), (3, –1), (6, –2), (9, –3), (12, –4)}

3. A third way to represent a function is with an algebraic equation that expresses a relationship between the independent variable, x, and the dependent variable, y. Particular pairings of the function can be found by evaluating the rule of the function at specific domain values. Describe the function for each algebraic rule. Then state the domain and the range of the function. Unless otherwise stated, the domain is the largest set of real numbers defined for the rule of the function.

    a. Function M: $y = -2x$ b. Function N: $y = \dfrac{1}{x}$ c. Function O: $y = x^3 + 1$

4. A graph can also be used to represent a function by plotting domain and range values on the Cartesian coordinate system. Describe the function in each graph. Then state the domain and the range values used in the function.

   a. Function *P*   b. Function *Q*   c. Function *R*

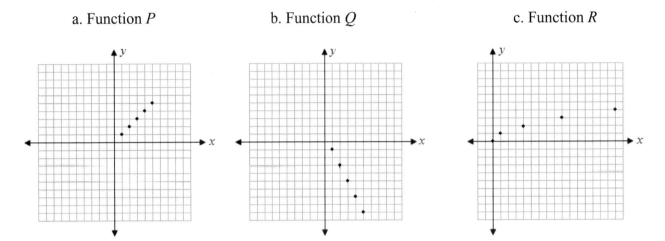

Part B: Classifying Functions

Functions can be classified in a number of ways. Consider the next several problems.

1. Functions can be classified as increasing, decreasing, or constant. A function is **increasing** if increasing values in the domain correspond with increasing values in the range. A function is **decreasing** if increasing values in the domain correspond with decreasing values in the range. A function is **constant** if increasing values in the domain correspond with the same value in the range. Based on the given information, classify the 12 functions in Part A as increasing, decreasing, or constant.

2. Functions can also be classified as linear or non-linear. A function is **linear** if it has a constant rate of change. That is, a consistent change in the domain values corresponds to a consistent change in the range values. Based on the given information, classify the 12 functions in Part A as linear or non-linear.

3. What other ways can you use the patterns of change you see to classify the 12 functions in Part A?

Part C: Summary and Connections

1. a. Which representation in Part A was most difficult for you to identify and describe the pattern of change represented by the function? Can you explain why?
   b. What implications might your answer to the previous question have in how you might represent functions were you to teach them in the classroom?

2. Can you think of five other patterns of change that can be represented with functions that were not demonstrated in this exploration?

Exploration 9.6 Functions in the Elementary Classroom

Purpose: Students explore two function activities that can be used in the elementary classroom.

Materials: Spinners, dice, and the cards for "Can you guess my rule?"

Part A: Function Spin

Elementary students can play a number of games that incorporate functional thinking. One such game is called "Function Spin". To play the game, a group of 3 or 4 students are given a die and spinner with several rules for functions. Spinners A and B show two suggestions, but others can be created. Play begins with a spin to determine the rule of the function. Each person then rolls the die and evaluates the result in the rule of the function. The rule of the function, the inputs, and the outputs should be recorded in a table. Play continues for two rounds, after which a new rule is selected from the spinner.

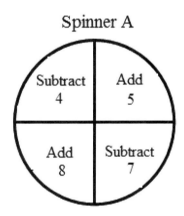
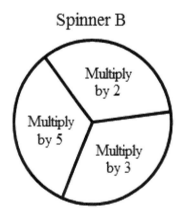

1. Form a small group with 2 or 3 of your peers and play three games of "What's My Rule?" You can use Spinner A, Spinner B, or make your own.

2. The game can be extended in several ways. Select one of the following extensions and play several games using the new rules.
    a. Apply the same rules given above but use two rules created by two spins from the same or different spinners. Use the appropriate order of operations.
    b. Work in pairs. One person hides the spinner before spinning. The other person must guess the rule by giving up to 3 inputs to which the "secret rule" is applied. The inputs and outputs can be recorded in a table.
    c. Form a small group and select a leader. The leader creates or chooses a spinner and shows it to the group. The leader then hides the spinner and spins twice. The other group members give three inputs. The leader does both operations and gives the outputs, recording the results in a table. The group must figure out both rules. (NOTE: There may be more than one combination of rules that works!)

Part B: Can You Guess My Rule?

Another game that incorporates functional thinking is the card game "Can You Guess My Rule?" The game is played with two people and uses the deck of 16 cards provided on the following page. The deck is shuffled and player A selects a card without showing it to Player B. Player B then gives Player A an input. Player A uses the input in the rule from the card and tells Player B the output. Player B records the input and output in a table. Play continues in this way until Player B guesses the rule. Player B may guess the rule at any time. If Player B's rule is correct, he scores one point for each number in the input column (e.g., one point for each guess). If Player B's rule is incorrect, he writes down another input and play continues. Players switch positions and play continues. The game consists of 3-5 rounds. The player with the **lowest** score wins because this player guessed the rules with the fewest number of clues.

1. Play a game of "Can You Guess My Rule?" with a partner.

2. Discuss with your partner different ways to extend this game.

Part C: Summary and Connections

1. Discuss with your group or partner how the games in this exploration make use of functional thinking. Write a paragraph or two summarizing your thoughts.

2. How could you modify the activities to allow for students with different abilities?

Cards for Can You Guess My Rule?

Can You Guess My Rule? Multiply input by 4.	**Can You Guess My Rule?** Add 8 to the input.
Can You Guess My Rule? Multiply the input by itself.	**Can You Guess My Rule?** Multiply the input by 3 and subtract 2.
Can You Guess My Rule? Multiply the input by 4 and subtract 3.	**Can You Guess My Rule?** Subtract 6 from the input.
Can You Guess My Rule? Add 2 to the input and multiply by 5.	**Can You Guess My Rule?** Multiply the input by itself and add 3.
Can You Guess My Rule? Multiply the input by 6 and subtract 1.	**Can You Guess My Rule?** Multiply the input by 1 more than the input.
Can You Guess My Rule? Multiply the input by 9 and add 4.	**Can You Guess My Rule?** Add 10 to the input.
Can You Guess My Rule? Subtract 7 from the input.	**Can You Guess My Rule?** Divide the input by 2 and subtract 1.
Can You Guess My Rule? Divide the input by 10.	**Can You Guess My Rule?** Multiply the input by 6.

Exploration 9.7 Graphing Functions on a Calculator

Purpose: Students explore the graphing features of a graphing calculator.

Materials: Graphing calculator, paper, and pencil

Part A: Graphing Functions

Graphing calculators are powerful learning tools because they enable students to graph complex functions quickly and efficiently.

1. Use a graphing calculator to graph each function. Sketch a picture of the graph that appears on your calculator.

    a. $y = (x - 3)^2$
    b. $y = x^3 + 4$
    c. $y = \sqrt{x + 4}$
    d. $y = \dfrac{3}{x - 4}$

2. Describe the general procedure you used to generate each graph.

Part B: Analyzing a Graph with the Trace Feature

After a graph has been created, it can be analyzed using features on a graphing calculator. One way to analyze a graph is to use the trace feature. Consider the following questions.

1. Graph the function $y = x^2 - 7$. Set the range on the window to the standard settings; that is, the minimum and maximum x- and y-values to -10 and 10 respectively.

2. Describe the graph. What particular points might be of interest on the graph?

3. Locate the trace feature on your calculator.
    a. What new items appear on your screen after bringing up the trace feature?
    b. What happens on your screen as you press the arrow buttons on your calculator?

4. Use the trace feature to answer each question.
    a. What approximate x-value gives the smallest y-value on the graph?
    b. What are the approximate x-values at which the graph crosses the x-axis?
    c. What y-value does the calculator give you when $x = 4$? Does this match the actual value of the function at $x = 4$? What might account for any discrepancies?

5. Repeat Questions 1 through 4 using the functions $y = x^3 - 5x - 2$.

Part C: Analyzing a Graph with the Zoom Features

Another way to analyze a graph is to use the zoom features. Consider the following tasks.

1. Graph the function $y = x^2 - x - 5$. Set the range on the window to the standard settings; that is, set the minimum and maximum x- and y-values to -10 and 10 respectively.

2. Describe the graph. What particular points might be of interest on the graph?

3. Locate the ZOOM features on your calculator. Explore each feature to see what it does. Reset the ranges on the window to the standard settings as necessary.

4. Select a zoom feature and use it to zoom in on one of the points where the graph crosses the x-axis. How did zooming in on this point change the range of values that are shown in the window?

5. Devise a strategy for using the zoom features to estimate the x-intercepts of the function $y = x^2 - x - 5$ to three decimal places. Write several sentences that describe your method.

6. Graph the function $y = x^3 - 5x - 2$. Use your strategy from Question 5 to estimate the x-intercepts of the function to three decimal places.

Part D: Summary and Connections

1. Explore other graphing features of your calculator. Which do you find most useful?

2. a. What role do you think calculators play in the elementary school? What about middle school?
   b. At what grade level do you think it would be appropriate for students to learn how to use a four-function or a fraction calculator? What about a graphing calculator?

Exploration 9.8 Exploring the Slope and *Y*-Intercept of a Line

Purpose: Students use dynamic geometry software to explore the slope and *y*-intercept of linear functions.

Materials: Dynamic geometry software

Part A: Slope and *Y*-Intercept of a Line

Because of its dynamic nature, geometry software can provide a powerful place to explore linear functions. Open the software and bring up a coordinate grid. After the coordinate grid appears, click on the *y*-axis and construct a point on it. Label the point *B*. Next, construct a point anywhere on the screen and label it *A*. Construct a line between points *A* and *B*. Label the line *l*. Click and drag on *A*. You should notice that *B* remains fixed on the *y*-axis as you move *A*.

1. Place *A* in the first quadrant and *B* at the origin. Highlight *l* and use the software to measure the slope of *l*. Next, have the software generate the equation of line *l*.
    a. What do you notice about the value of the slope and the equation of line *l*? What form is the equation in?
    b. What happens to the slope as you drag *A* up in the first quadrant? Down in the first quadrant?
    c. What happens to the slope of line *l* when you place *A* in the
        i. Second quadrant?   ii. Third quadrant?   iii. Fourth quadrant?
    d. What happens when *A* is placed on the *x*-axis? Can you explain this?
    e. What happens when *A* is placed on the *y*-axis? Can you explain this?

2. Place *A* back in the first quadrant. Highlight point *B* and use the software to put its coordinates on the screen. Click and drag *B*.
    a. What happens to the *x*-coordinate of *B* as it moves?
    b. What happens to the *y*-coordinate of *B* as it moves? How does this relate to the equation of line *l*?
    c. What happens to the slope of line *l* when *B* is below *A*? Above *A*? Can you explain this?

Part B: Parallel and Perpendicular Lines

Place *A* back in the first quadrant. Construct a new point on the *y*-axis and label it *C*. Drag *C* so that it is above *B*.

1. Use the software to construct a line parallel to *l* that passes through *C*. Label the new line *k*. Drag point *A*. You should see that lines *l* and *k* remain parallel. Highlight line *k* and use the software to measure its slope and to give its equation.
    a. How are the equations for lines *l* and *k* the same? How are they different?

b. What do you notice about the slopes of lines *l* and *k*? Based on what you see, write a conjecture about the slopes of parallel lines.

c. Test your conjecture by dragging *A* to different locations. What do you notice about the slopes of lines *l* and *k*? Does this confirm your conjecture?

2. Highlight and delete line k. Use the software to construct a line perpendicular to *l* that passes through *C*. Label the new line *n*. Drag point *A*. You should see that lines *l* and *n* remain perpendicular. Highlight line *n* and use the software to measure its slope and to give its equation.

    a. How are the equations for lines *l* and *n* the same? How are they different?

    b. Highlight the slopes of lines *l* and *n*. Use the software to compute the product of the two slopes. What is the value of this product? Based on what you see, write a conjecture about the slopes of perpendicular lines.

    c. Test your conjecture by dragging *A* to different locations. What do you notice about the slopes of lines *l* and *n*? Does this confirm your conjecture?

Part C: Summary and Connections

1. Write a summary of what you learned about the slopes and *y*-intercepts of parallel and perpendicular lines.

2. What is the value in using geometry software to learn about lines?

Exploration 9.9 Arithmetic Sequences and Linear Functions

Purpose: Students use different representations to connect arithmetic sequences to linear functions.

Materials: Paper and pencil

Part A: Arithmetic Sequence and Linear Functions

Arithmetic sequences are closely connected to linear functions. To better understand the connection, consider the following questions.

1. Complete the table by finding the first eight terms of the arithmetic sequence 3, 7, 11, ...

Term	1st	2nd	3rd	4th	5th	6th	7th	8th
Value	3	7	11					

2. A linear function exhibits a pattern of constant change; that is, a consistent change in one set of numbers corresponds to a consistent change in another.
    a. Consider the table in Question 1. Why does the table exhibit a function?
    b. What are the independent and dependent variables of the function?
    c. Describe the pattern that exists between the positions of the terms and their values. Is it a pattern of constant change?
    d. What does your answer to the previous question imply about the sequence?

3. The table in Question 1 exhibits a function. Graph the values from the table.
    a. What do you notice about the graph? Does this confirm any observation you made in Question 2?
    b. What is the domain and range of this function?
    c. What is the domain of any arithmetic sequence when the sequence is viewed as a function?

4. Find the general term of the sequence 3, 7, 11, ...
    a. Why does the general term represent a linear equation?
    b. The common difference is the coefficient of the variable in the general term. What does this coefficient represent in the corresponding linear function?

5. Repeat Questions 1 through 4 using the arithmetic sequence 8, 5, 2, ...

6. The first four questions illustrate that every arithmetic sequence is a linear function. Consequently, every linear function must be associated with an arithmetic sequence. For instance, consider the linear function $y = 2x + 5$.
   a. How can the terms of an arithmetic sequence be generated from the equation of the function? Use your method to generate the first five terms of the arithmetic sequence. Identify the first term and the common difference.
   b. Find the formula for the general term of the sequence. How does it compare to the formula of the function?

7. a. Suppose you were given a linear function. How could you use the algebraic rule of the function to find the first term and the common difference of the corresponding arithmetic sequence without having to evaluate the function at any values?
   b. Use your method to find the first term and the common difference of the arithmetic sequences associated with each linear function.

   i. $y = 3x - 5$ ii. $y = -2x + 4$ iii. $y = \dfrac{1}{2}x + \dfrac{3}{4}$

Part B: Summary and Connections

1. Write a paragraph or two summarizing what you have learned about arithmetic sequences and linear functions.

2. Geometric sequences are also connected to a particular type of function. What is it?

Exploration 9.10 Solving Linear Equations with Algebra Tiles

Purpose: Students use algebra tiles to represent and solve linear equations.

Materials: Algebra tiles and an equation mat

Part A: Representing Algebraic Expressions with Algebra Tiles

Algebra tiles are a manipulative that uses square and rectangular tiles of different colors to represent algebraic expressions. This exploration uses two sizes of tiles: squares and rectangles. Square tiles are called unit tiles. The yellow square has a value of 1 and the red square a value of (–1). Rectangular tiles are called *x*-tiles. The yellow rectangle has an unknown value of *x* and the red rectangle has value of (–*x*), or the opposite of *x*.

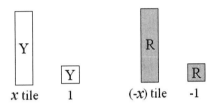

1. To represent an algebraic expression, select the appropriate number and color of each type of tile. For instance, the expression $3x + (-4)$ is represented with three yellow rectangles and 4 red squares. Write the algebraic expression that is represented in each diagram.

   a.   b. 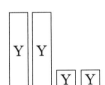   c.   d.

2. Any time two unit tiles or two *x*-tiles of different colors are placed together, they have a net value of zero and are called a **zero pair**. Write the algebraic expression that is represented in each diagram. Take all zero pairs into account before writing the expression.

   a. 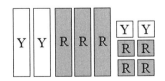   b. 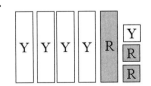   c.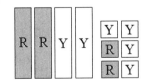

3. Draw a diagram that shows how to use algebra tiles to represent each expression with the fewest tiles possible.
   a. 3   b. $-3x + 1$ c. $4x + 4$ d. $3x + (-6)$

4. Draw a diagram that shows how to use algebra tiles to represent each expression in three ways.

   a. 2 b. $-4x$ c. $-x + 3$ d. $2x + 3$

5. Algebraic equations can be represented with algebra tiles by using an equation mat. An equation is represented by placing the appropriate tiles on either side of the vertical line, which is used to represent the equals sign. What linear equation is represented on each mat?

   a.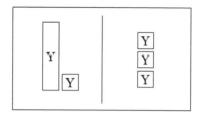
   b.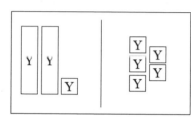
   c.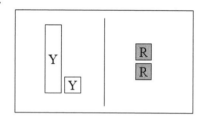
   d.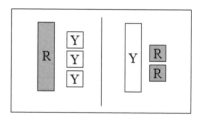

6. Draw a diagram that shows how to use algebra tiles and an equation mat to represent each equation.

   a. $-x + (-4) = 2$
   b. $2x + 1 = 5$
   c. $3x + 3 = -2x + (-2)$
   d. $-3x + (-2) = x + 4$

Part B: Solving Linear Equations with Algebra Tiles

If algebra tiles are placed on an equation mat, they can also be used to solve linear equations. To do so, represent the equation with the appropriate tiles and then use zero pairs to remove tiles from the equation mat until a positive x-tile is on one side and a set of unit tiles on the other. For instance, consider the equation $2x + 3 = -5$. After representing it on an equation mat, we can remove the three positive unit tiles from the left-hand side by placing 3 negative unit tiles on both sides of the equation to maintain the balance.

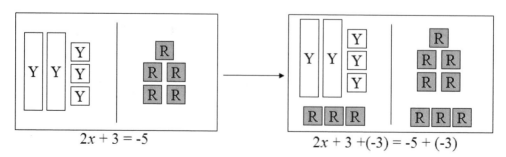

After removing the zero pairs, 2 positive x-tiles are left on the right and 8 negative unit tiles are on the left.

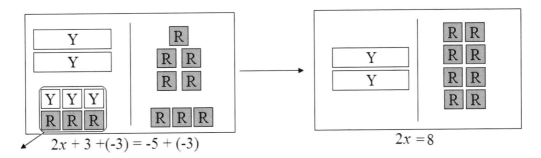

From this point, we can see that each positive x-tile can be paired with 4 negative unit tiles, so we conclude that $x = -4$.

1. Draw a diagram that shows how to use algebra tiles and an equation mat to represent and solve each equation. Record the algebraic representation for each step in your solution.
   a. $x + 3 = 8$
   b. $5 + x = -2$
   c. $3x + 5 = 8$
   d. $x + (-3) = 2x + 4$

2. Give an example of a linear equation that cannot be solved easily with algebra tiles. Explain why it is difficult.

Part C: Summary and Connections

1. What are the advantages and disadvantages of using algebra tiles when introducing students to solving linear equations?

2. Why is it important to record the algebraic representations as you solve linear equations with algebra tiles?

Exploration 9.11 Student Errors in Solving Linear Equations and Inequalities

Purpose: Students examine common errors that middle school students make when solving linear equations and inequalities.

Materials: Paper and pencil

Part A: Student Errors in Solving Equations and Inequalities

As middle school students learn to solve simple equations and inequalities, it is not unusual for them to make mistakes. As a teacher, it is up to you to diagnose their errors and offer feedback to help them correct their thinking. Examine each equation or inequality. Identify the mistake and explain how you might correct the student's error.

1. Tamarcus:

    $3x + 4 = 11$
    $3x = 15$
    $x = 5$

2. Ibrahim:

    $2(x+1) = 5$
    $2x + 1 = 5$
    $2x = 4$
    $x = 2$

3. LaTasha:

    $2x + 3x + 4 = 4x + 13$
    $9x + 4 = 13$
    $9x = 9$
    $x = 1$

4. Macy:

    $6x - 7 = 12$
    $6x = 12$
    $x = 2$

5. Devon:

    $-3x - 8 < 7$
    $-3x < 15$
    $x < -5$

6. Bridget:

    $4x - 7 \geq 17$
    $4x \leq 24$
    $x \leq 6$

Part B: Summary and Connections

1. Other than the ones presented here, what other errors do you think students might make as they learn to solve simple equations and inequalities?

2. It is very common for children to solve simple one-step equations as early as first or second grade. In what context is this likely to occur?

Exploration 9.12 Mathematical Modeling and Functions

Purpose: Students use different representations to model situations involving functions.

Materials: Graph paper and pencil

Part A: Modeling with Functions

Because many real-world situations exhibit a pattern of change between two variables, we can use functions and their representations to model them. For the following situations, use functions to express the relationship between the independent and dependent variables.

1. The managers of a widget factory want to track the daily cost of running the factory. They know the factory has a fixed cost of $50,000 per day and each widget costs $10 to produce.
    a. What are the two unknowns in this situation? Which one is the independent variable? The dependent variable?
    b. Why is the relationship between the two variables a functional relationship?
    c. Write an equation to represent the function between the total daily cost, c, and the number of widgets produced, w. What is the domain and range of the function?
    d. What is the value of the function at $w = 3,000$? What does this value represent?
    e. Suppose the factory has a budget of $230,000 for its total daily cost. What is the maximum number of widgets that can be produced?
    f. Represent the function between c and w with a table of values and use the values to draw a graph.
    g. Where does the graph cross the horizontal axis? The vertical axis? Interpret the coordinates of these points in terms of the context of the problem.

2. Sergio borrowed $7,200 from his father to buy a used car. His father is not charging him interest on the loan, but expects him to repay $200 per month. Sergio wants a way to keep track of how much he owes his father on a monthly basis.
    a. What are the two unknowns in this situation? Which one is the independent variable? The dependent variable?
    b. Why is the relationship between the two variables a functional relationship?
    c. Write an equation to represent the function between the amount of money Sergio owes, a, and the number of months, m, he has made a payment on the loan. What is the domain and range of the function?
    d. How much does Sergio owe after he has made payments for 20 months?
    e. How long will it take Sergio to repay half of the loan?
    f. Represent the function between a and m with a table of values and use the values to draw a graph.
    g. Where does the graph cross the horizontal axis? The vertical axis? Interpret the coordinates of these points in terms of the context of the problem.

Part B: Writing Story Problems Involving Functions

As a teacher, it is often necessary to write a story problem that exemplifies a particular mathematical concept. Write a story problem for each functional situation.

1. Let t represent the number of tickets sold to a play and let a represent the amount of money made by the play on a given night. Write a story problem relating these two variables so that $a = 15t + 200$.

2. Let t represent a number of hours and let d represent a distance traveled. Write a story problem relating these two variables so that $d = 150 - 45t$.

3. Write a story problem that uses the function $y = \frac{2}{3}x$. Be sure to identify what the variables x and y represent in the story problem.

Part C: Summary and Connections

1. Can you think of a real-world situation that exhibits a functional relationship but cannot be represented with an algebraic equation?

2. Why is it valuable to identify the domain and range of functions that model real-world situations?

Exploration 9.13 Mathematical Modeling and Equations

Purpose: Students use equations to model and solve problems in different situations.

Materials: Paper and pencil

Part A: Modeling with Equations

Identify the unknowns and then write an equation to represent the situation in each problem. Solve the equation and give the answer in terms of the original problem.

1. A school sold 520 tickets to a spring concert. They sold 84 more adult tickets than student tickets. How many of each kind of ticket were sold?

2. On her first three 50-point tests, Sheila got a 38, a 46, and a 48. If she wants at least a 90% average in the class, what is the lowest score she can get on her last 50-point test?

3. Jill was charged $137 for repairs to her TV. The bill included labor and $32 for parts. If the repairman charges $30 per hour for labor, how many hours did he work on the TV?

4. A rectangle's width is 4 times its length. The perimeter is 30 feet. What is the length and width of the rectangle?

Part B: Writing Story Problems Involving Equations

As a teacher, it is often necessary to write story problems that exemplify a particular concept. Write a story problem that involves solving an equation for each situation.

1. Write a ticket problem similar to the one in Part A Question 1, in which the total number of tickets is known and there is a relationship between the number of student tickets and the number of adult tickets.

2. Write an average problem similar to the one in Part A Question 2. Use a different number of tests and a different average that the student wants to achieve.

Part C: Summary and Connections

1. Describe a situation in which you have used an algebraic equation to solve a problem that you encountered in your daily life. Why did you use an equation to solve the problem?

2. Do you think it makes sense to use contextual situations when students begin to solve equations? Why or why not?

Chapter 10
Geometrical Shapes

Chapter 10 marks a shift away from a study of number structures and algebraic thinking toward geometry. The word "geometry" comes from two Greek words that mean "earth measure," so it is literally the study of shapes, their properties, and their measures. We use it extensively, not only as we interact and describe the world around us, but also as we learn other mathematical topics and academic disciplines.

Geometry covers a wide variety of topics, so in this chapter, we focus on analyzing shapes and their properties. In the elementary grades, students first learn to name and classify two- and three-dimensional shapes by visual means. They begin to analyze the shapes for their defining characteristics and their properties. As students learn geometry, they also develop their spatial reasoning skills. **Spatial reasoning,** or **spatial sense**, refers to the ability to understand and operate on the relative position of objects in space with respect to a particular position. It also includes the ability to comprehend and perform imagined movements on objects in two- and three-dimensional space. Spatial reasoning skills are often developed through navigation activities, art projects, and other lessons in which students physically manipulate objects or view them from a variety of perspectives.

The explorations in Chapter 10 analyze shapes and their properties using a variety of representations and manipulatives. Specifically,

- Explorations 10.1 - 10.3 use manipulatives and turtle geometry to focus on lines, planes, and angles.
- Explorations 10.4 - 10.7 explore the characteristics and properties of triangles.
- Explorations 10.8 - 10.11 focus on the characteristics and properties of quadrilaterals and other polygons.
- Explorations 10.12 - 10.15 analyze the properties of three-dimensional shapes.

Exploration 10.1 Measuring Angles with Manipulative Pieces

Purpose: Students find the measures of the angles in manipulative pieces and then use them to measure other angles.

Materials: Pattern blocks, tangrams, and fraction circles

Part A: Angles in Manipulative Pieces

1. Trace each shape from a set of the pattern blocks onto notebook paper. Find the measure of each interior angle of each pattern block using only the fact that there are 360° in a complete rotation of a ray about its endpoint. Write the measure of each angle on the pattern block. Explain the process you used to find the angle measures.

2. Trace each shape from a set of tangram pieces on to notebook paper. Find the measure of each interior angle of each shape using only the fact that there are 360° in a complete rotation of a ray about its endpoint. Write the measure of each angle on the tangram piece. Explain the process you used to find the angle measures.

3. Find the measure of the angle for each different size piece in a set of fraction circles. Write the measure of the angle on the piece. Explain the process you used to find the angle measures.

Part B: Measuring Angles

1. Use manipulative pieces to approximate the measure of each angle. What pieces did you use and why did you choose them?

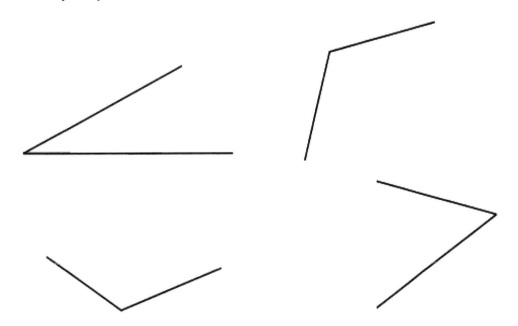

2. Use manipulative pieces to approximate the measure of each interior angle of the given shape. What pieces did you use and why did you choose them?

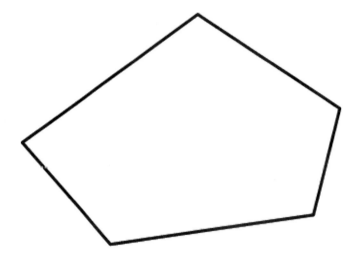

Part C: Creating Angles

1. Use manipulative pieces to draw angles with the following measures. Explain how you created each angle.
    a. 15°   b. 135°   c. 165°   d. 105°

2. What other angle measures can you create using the manipulative pieces?

Part D: Summary and Connections

1. Write a short summary about what you have learned about angle measures by working through the activities in this lab.

2. How can you use the fact that there are 360° in a complete rotation of a ray about its endpoint to visualize an angle of 1°?

Exploration 10.2 Segments and Angles on a Geoboard

Purpose: Students explore segments, angles, and lines on a geoboard.

Materials: A 6 × 6 geoboard and rubber bands

Part A: Segments on a Geoboard

Use a 6 × 6 geoboard to answer the following questions about segments and lines.

1. Use rubber bands to make the segments shown in the diagram.
    a. Which segments appear to be the same length? How can you be sure?
    b. How many different lengths are shown? Name them from shortest to longest.

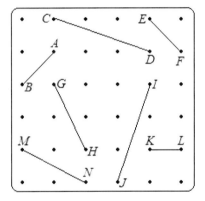

2. Continue the pattern shown on each geoboard until no more segments can be made. How many segments did you make and what are their lengths?
    a.
    b.

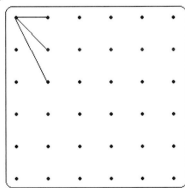   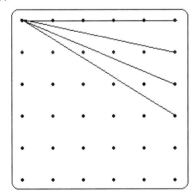

3. How many different lengths of line segments can be made on a 6 × 6 geoboard?

4. If possible, make a segment of each length.
    a. 3 units
    b. $\sqrt{5}$ units
    c. $\sqrt{13}$ units
    d. $\sqrt{18}$ units

5. Use rubber bands to make the segments shown in the diagram. Next, create each segment and describe how you know the segment meets the given conditions.
    a. Make a segment that is parallel to $\overline{AB}$ and goes through D.
    b. Make a segment that is perpendicular to $\overline{CD}$ and goes through F.
    c. Make a segment that is parallel to $\overline{EF}$ and goes through B.
    d. Make a segment that is perpendicular to $\overline{EF}$ and goes through C.

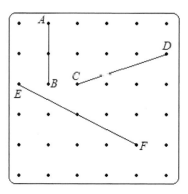

Part B: Angles on a Geoboard

1. Use rubber bands to create each type of angle.
    a. Acute   b. Right   c. Obtuse   d. Straight

2. If possible, use rubber bands to create an angle of each measure.
    a. 90°   b. 45°   c. 135°   d. 30°

3. On the bottom row of your geoboard, select the peg that is third from the left to serve as the vertex of an angle. Using the pegs directly to the right as the initial side of an angle, form as many angles as you can. Three possible angles are shown.
    a. How many angles of different measures can be made?
    b. How many of the angles are acute?
    c. How many of the angles are obtuse?
    d. How many angles are neither acute nor obtuse?

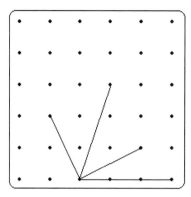

4. Repeat the previous activity, only this time, select the peg on the bottom row that is fourth from the left to serve as the vertex of an angle.
    a. How many angles of different measures can be made?
    b. How many of the angles are acute?
    c. How many of the angles are obtuse?
    d. How many angles are neither acute nor obtuse?
    e. How do your answers in this question compare to your answers in Question 3? Can you explain any similarities or differences?

Part C: Summary and Connections

1. What are the advantages and disadvantages of using a geoboard to teach concepts about segments and angles?

2. How could you use a geoboard to teach the idea of the slope of a line? What about the slopes of parallel and perpendicular lines?

Exploration 10.3 Turtle Geometry

Purpose: Students use segments and angles to trace and describe paths for a turtle.

Materials: Ruler, protractor, unlined paper, and a plastic turtle, bear, or other "creature" to help with perspective

Part A: Drawing the Path of a Turtle

Each of the following paths indicates the movements of an imaginary turtle. Trace the path of the turtle on a sheet of unlined paper. Begin by placing a starting point near the center of the paper and marking it with START. Use a ruler and protractor to draw the path of the turtle. Mark the ending point of the path with STOP.

Path 1
  Forward 3 cm
  Right 90°
  Forward 2 cm
  Right 45°
  Forward 5 cm

Path 2
  Forward 4 cm
  Left 45°
  Forward 5 cm
  Right 90°
  Forward 6 cm

Path 3
  Forward 3 cm
  Right 120°
  Forward 7 cm
  Left 90°
  Forward 6 cm

Path 4
  Forward 2 in.
  Right 90°
  Forward $1\frac{1}{2}$ in.
  Left 50°
  Forward $2\frac{1}{4}$ in.
  Right 130°
  Forward $1\frac{3}{4}$ in.

Part B: Describing the Path of a Turtle

1. a. Write a set of directions that will guide a turtle along the following path if he begins at point *A*, faces the top of the page, and travels in a clockwise direction. Use the given lengths.
   b. Write a set of directions that will guide a turtle along the following path if he begins at point *B*, faces the left side of the page, and travels in a counterclockwise direction. Use the given lengths.

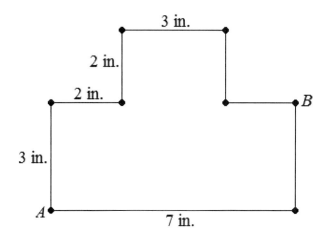

2. Write a set of directions that will guide a turtle along the following path if he begins at point *C*, faces the top of the page, and travels in a counter clockwise direction. Use the given lengths and angle measures.

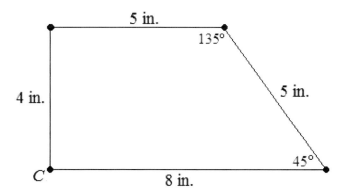

Part C: Summary and Connections

1. a. Create your own set of directions for a turtle path. Give them to a classmate to draw a picture of the path.
   b. Create your own picture of a turtle path. Give them to a classmate to write a set of directions that describes the path.

2. Explore Turtle Geometry further at the National Library of Virtual Manipulatives at http://nlvm.usu.edu/en/nav/vlibrary.html

Exploration 10.4 Sorting Planar Shapes

Purpose: Students use common characteristics to sort planar shapes.

Materials: Shape sorting cards and scissors

Part A: Sorting Planar Shapes

Planar shapes can be sorted in a number of ways. Cut out the 22 shape sorting cards and sort them as follows.

1. Look for common characteristics among the cards and sort them in five or six ways using these common characteristics. Make a list of the characteristics you used to sort the cards. Share them with two or three of your peers. How many different ways were you and your peers able to sort the cards?

2. In geometry, we are interested in several sets of planar shapes. The broadest of these sets is the set of curves. A **curve** is any figure that can be traced without lifting the pencil from the paper. Sort the shape cards into curves and non-curves.

3. Consider only those shapes that are curves. A **closed** curve is any curve that begins and ends at the same point. Sort the curves into closed curves and non-closed curves.

4. Consider only those shapes that are curves. A **simple** curve is any curve that never crosses itself except at the point where the curve begins and ends. Sort the curves into simple curves and non-simple curves.

5. Consider only those shapes that are simple, closed curves. Simple, closed curves can be convex or concave. A simple, closed curve is **convex** if for any two points inside of the curve, the segment connecting the points lies entirely in the interior of the curve. A simple, closed curve is **concave** if it is not convex.
    a. Draw a diagram of a convex curve. Use it to explain to one of your peers what it means for a simple, closed curve to be convex.
    b. Sort the shapes that are simple, closed curves into convex and concave curves.

6. Consider only those shapes that are simple, closed curves. A **polygon** is a simple, closed curve made entirely of line segments.
    a. Sort the simple, closed curves into polygons and non-polygons.
    b. Are some polygons convex and others concave? Explain.

7. If possible, draw an example of a curve that is
    a. Simple and closed.
    c. Non-simple and non-closed.
    e. Concave and non-simple.
    b. A non-closed polygon.
    d. A convex polygon.
    e. A non-simple polygon.

Part B: Summary and Connections

1. Why can sorting shapes be useful in learning the attributes of different shapes?

2. How did this sorting activity help you learn the definitions of some geometric terms?

3. Write a paragraph or two that explains how you might use a sorting activity to teach children about the different types of triangles or quadrilaterals.

Shape Sorting Cards

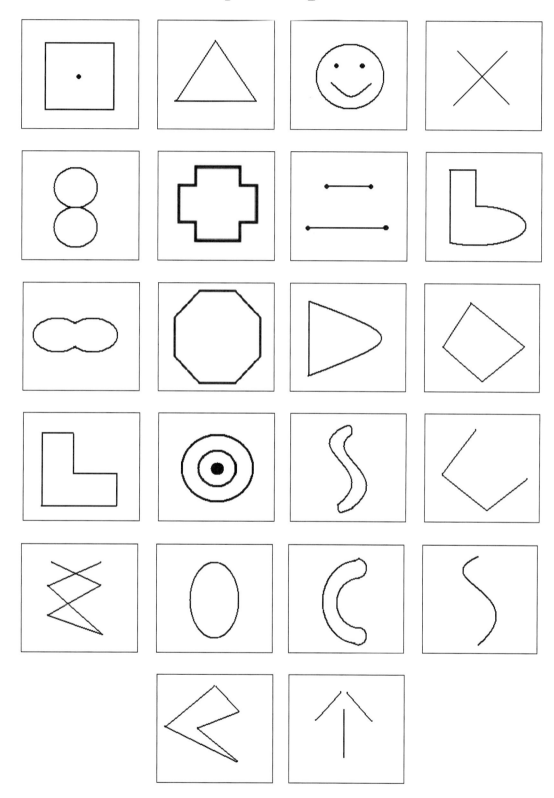

Exploration 10.5 The Sides and Angles of a Triangle

Purpose: Students use paper triangles to discover properties of triangles.

Materials: Ruler, protractor, and unlined paper

Part A: Sides of a Triangle

Use a ruler to draw three triangles on three sheets of paper. Make one triangle acute, one right, and one obtuse. Try to make each of the sides at least 4 or 5 inches long. Label the angles in each triangle ∠1, ∠2, and ∠3. Label the sides in each triangle a, b, and c so that a is opposite ∠1, b is opposite ∠2, and c is opposite ∠3. Measure the length of each side in each triangle and record the measures in the given space.

Acute Triangle
    Length of a = _____   Length of b = _____   Length of c = _____

Right Triangle
    Length of a = _____   Length of b = _____   Length of c = _____

Obtuse Triangle
    Length of a = _____   Length of b = _____   Length of c = _____

1. Compare each side to the angle that is opposite of it. What do you notice about the length of the side compared to size of the opposite angle?

2. In each triangle, add the lengths of any two sides. Compare the sum to the length of the third, unused side. What do you observe?

Part B: Angles of a Triangle

Use the same triangles from Part A. Use a protractor to measure each angle in each triangle. Record the measures below.

Acute Triangle
    $m\angle 1$ = _____ $m\angle 2$ = _____ $m\angle 3$ = _____

Right Triangle
    $m\angle 1$ = _____ $m\angle 2$ = _____ $m\angle 3$ = _____

Obtuse Triangle
    $m\angle 1$ = _____ $m\angle 2$ = _____ $m\angle 3$ = _____

1. Compare each angle to the side that is opposite of it. What do you notice about the size of the angle compared to the length of the opposite side? Why does this make sense in light of your answer to Question 1 in Part A?

2. a. What value do you get if you add the measures of the three angles in each triangle?
   b. Compare your answer to the previous question to the answers obtained by some of your peers. What do you observe? What conjecture can you make about the sum of the angles in any triangle?

Part C: Isosceles Triangles

1. △ABC and △DEF are isosceles triangles. Measure each angle and each side in each triangle. Use your measures to answer the following questions.
   a. What is the defining characteristic of an isosceles triangle?
   b. Name the pair of congruent sides in each triangle.
   c. Two angles in each triangle are congruent. What is the relationship between the congruent angles and the congruent sides?
   d. Complete each statement by filling in the blanks.
      i. If two sides of a triangle are congruent, then the angles opposite them are

      _____ and the triangle is _____.

      ii. If two angles of a triangle are congruent, then the sides opposite them are

      _____ and the triangle is _____.

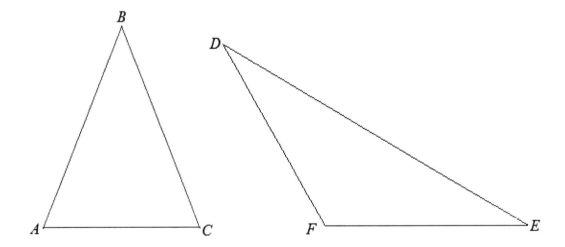

2. △GHI is an equilateral triangle. Measure each angle and each side. Use your measures to answer the following questions.
   a. What is the defining characteristic of an equilateral triangle?
   b. Name the congruent sides in the triangle.
   c. The three angles in the triangle are congruent. What is the relationship between the congruent angles and the congruent sides?

d. Complete each statement.
   i. If a triangle is equiangular, then it is also _____.

   ii. If a triangle is equilateral, then it is also _____.

e. Refer to the definition of an isosceles triangle. Is an equilateral triangle isosceles? Why or why not?

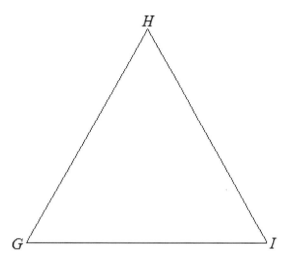

Part D: Summary and Connections

1. Write a short paragraph that summarizes the properties of triangles that you learned in this exploration.

2. What role did measurement play in identifying the properties of triangles? What role did inductive reasoning play?

3. How might you adapt this exploration for use in the elementary grades?

Exploration 10.6 Angle Measures in Triangles

Purpose: Students discover two properties of angle measures in triangles.

Materials: Ruler, protractor, 3 different colored pencils or markers, paper, and scissors

Part A: The Sum of the Interior Angles

Use a ruler to draw three triangles on three separate sheets of paper. Make one triangle acute, one right, and one obtuse. Be sure to make each of the sides at least 4 or 5 inches long. Label the angles ∠1, ∠2, and ∠3, and label the sides *a*, *b*, and *c* so that side *a* is opposite ∠1, *b* is opposite ∠2, and *c* is opposite ∠3. Use a protractor to measure the angles in each triangle. Record the measures in the given spaces. Cut out each triangle and trace it onto another sheet of paper. Be sure to trace the triangle in the center of the paper so that the sides can be extended in the next activity.

Acute Triangle
$m\angle 1 =$ _____ $m\angle 2 =$ _____ $m\angle 3 =$ _____

Right Triangle
$m\angle 1 =$ _____ $m\angle 2 =$ _____ $m\angle 3 =$ _____

Obtuse Triangle
$m\angle 1 =$ _____ $m\angle 2 =$ _____ $m\angle 3 =$ _____

1. Add the measures of the three angles in each triangle. Write a conjecture about what you observe.

2. Color the interior of each angle with a different colored pencil. Tear off the corners of each triangle. Place the three angles so they share a common vertex. What do you notice? How does this support your conclusion in Question 1?

3. The two preceding problems show that the sum of the measures of the interior angles in a triangle is equal to 180°. Use the following diagram to write an argument that proves this fact. Assume that line *p* is parallel to $\overline{AB}$.

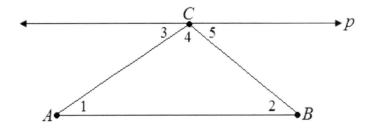

Part B: Exterior and Remote Interior Angles

Take the tracings of the three triangles you made in Part A and use a ruler to extend each of the sides to form the exterior angles of the triangles. An **exterior angle** is formed by the side of a triangle and the extension of an adjacent side. Use a protractor to measure the exterior angles in each triangle. Record the measures on the tracings.

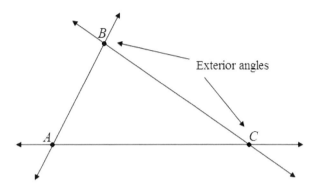

1. Compare the measure of an exterior angle at one vertex to the interior angles at the other two vertices. These interior angles are called remote interior angles to this exterior angle.
    a. Why does the phrase "remote interior angles" make sense?
    b. How does the measure of the exterior angle compare to the measures of its remote interior angles?
    c. Does your observation hold for the other exterior angles? What about the exterior angles in the other triangles?
    d. Write a conjecture about the relationship between the measure of an exterior angle of a triangle and the measures of its remote interior angles.

2. Find the corners that you tore off of the triangles in Part A. Take the corners of two interior angles and place them over the appropriate exterior angle on the tracing of the triangle. What do you notice? Does this support your observations from Question 1? Repeat the comparison using other angles from other triangles.

3. The two preceding problems show that the measure of an exterior angle in a triangle is equal to the sum of the measures of the two remote interior angles. Use your result from Part A to write an argument that proves this fact.

Part C: Summary and Connections

1. How can the sum of the interior angles in a triangle be used to find the sum of the interior angles in a quadrilateral? What about a pentagon?

2. How did your arguments in this exploration mimic the procedures used to discover the facts about the angles in a triangle?

Exploration 10.7 Making Sense of the Pythagorean Theorem

Purpose: Students use square tiles to make sense of the Pythagorean Theorem and one of its proofs.

Materials: 1-in. square tiles, 1-in. grid paper, paper, and scissors

Part A: Making Sense of the Pythagorean Theorem

In the center of a sheet of grid paper, draw a right triangle with legs of lengths 3 in. and 4 in. Form the right angle by using the grid lines on the paper. Next, use 1-in. square tiles to build a square on each leg of the right triangle. Let Square A be the square with sides of length 3 in. and Square B be the square with sides of length 4 in.

1. a. What is the area of Square A? The area of Square B?
   b. What is the connection between the length of the side of the square and its area?
   c. How can you express the area of a square in terms of the side of the triangle on which it is built?

2. Combine the tiles that made Square A with the tiles that made Square B.
   a. Can you use these tiles and only these tiles to build a square on the hypotenuse of the triangle?
   b. If such a square is possible, what is its area? How do you know?
   c. How can you use the area of the square to determine the length of the hypotenuse on which this square is built?
   d. How can you express the area of this square in terms of the length of the hypotenuse?

3. Repeat Questions 1 and 2. This time, draw a right triangle with legs of lengths 2 in. and 3 in. in the center of a sheet of grid paper. Use 1-in. square tiles to build a square on each leg of the right triangle. Let Square C be the square with sides of length 2 in. and let Square D be the square with sides of length 3 in. When answering Question 2 again, it may help to replace some plastic tiles with paper 1-in. tiles that can be torn.

4. Now consider a right triangle that has legs of lengths a and b and has a hypotenuse of length c. Use your observations from Questions 1, 2, and 3 to answer the following.
   a. If you built a square on the leg with length a, what would be its area?
   b. If you built a square on the leg with length b, what would be its area?
   c. If you built a square on the hypotenuse of the right triangle, how would its area compare to the areas of the squares on the legs?
   d. Write an equation to express the relationship among the areas of the three squares in terms of a, b, and c. What does the equation mean? Why does the equation make sense?

Part B: A Visual Proof of the Pythagorean Theorem

Get together with two of your peers. Assign one of the following diagrams to each person in the group. Each diagram contains two shapes; a right triangle with squares on its sides and a square labeled *WXYZ*. Have each person make four copies of their right triangle and one copy of each square on the sides of the triangle on another sheet of paper. Cut out the triangles and squares. Show that square *WXYZ* can be covered with the hypotenuse square and the four triangles. Next, show square *WXYZ* can be covered with the two leg squares and the four triangles. What can your group conclude from manipulating the figures in the diagrams? Can you explain why?

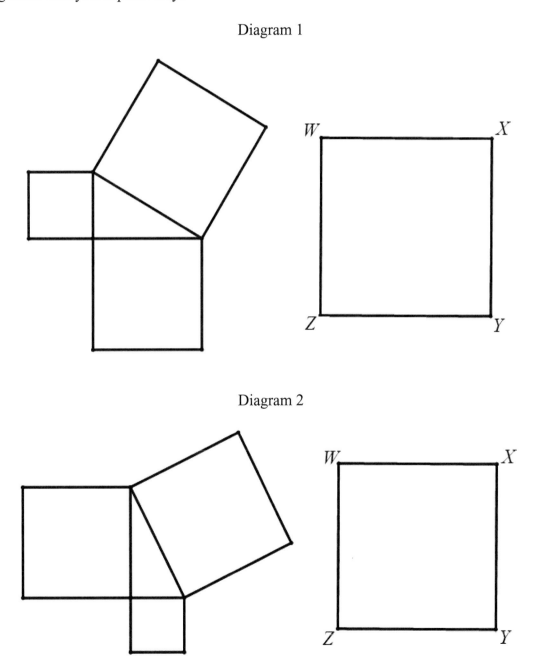

Diagram 1

Diagram 2

Diagram 3

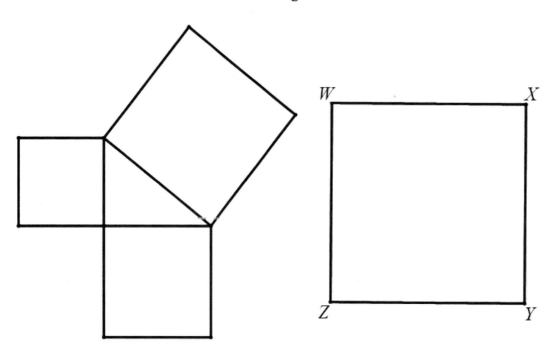

Part C: Summary and Connections

1. Summarize what you have learned about the area of squares and a right triangle.

2. A student says, "The Pythagorean Theorem is $a^2 + b^2 = c^2$." What is wrong with this statement?

3. James Garfield, the twentieth president of the United States, gave an original proof of the Pythagorean Theorem. Search the Internet to find his proof. Write a short paragraph that describes his method.

Exploration 10.8 Properties of Quadrilaterals

Purpose: Students use measurement to discover properties of quadrilaterals.

Materials: A set of quadrilaterals including parallelograms, rhombuses, rectangles, squares, and isosceles trapezoids, a ruler, and a protractor,

Part A: Properties of Parallelograms

A **parallelogram** is a quadrilateral that has both pairs of opposite sides parallel. Find the parallelograms in your set of quadrilaterals. Use a protractor to measure their angles and a ruler to measure their sides. Record the measures.

1. Two angles in a quadrilateral that do not share a common side are called **opposite angles**. Compare the measures of each pair of opposite angles in each parallelogram. Write a conjecture about what you observe. Does your conjecture hold for all parallelograms? How do you know?

2. Two angles in a quadrilateral that share a common side are called **consecutive angles.** Compare the measures of each pair of consecutive angles in each parallelogram. Write a conjecture about what you observe. Does your conjecture hold for all parallelograms? How do you know?

3. Two sides in a quadrilateral that do not share a common vertex are called **opposite sides**. Compare the measures of each pair of opposite sides in each parallelogram. Write a conjecture about what you observe. Does your conjecture hold true for all parallelograms? How do you know?

4. Draw the two diagonals in each parallelogram. Label the point of intersection P. Measure the length of each segment from each of the vertices to P.
   a. What is P in relationship to each diagonal of the parallelogram?
   b. What does this mean the two diagonals do to each other?

Part B: Properties of Rhombuses

A **rhombus** is a parallelogram in which all the sides are the same length. Find the rhombuses in your set of quadrilaterals and use them to answer each question.

1. Because a rhombus is a parallelogram, what must be true about its angles, sides, and diagonals? Check your conjectures by using a protractor to measure the angles and a ruler to measure the sides.

2. Draw the diagonals in each rhombus. Label the point of intersection P. Use a protractor to measure the four new angles made by the intersecting diagonals. Compare the measures of each angle at P. Write a conjecture about what you observe. Does your conjecture hold true for every rhombus? How do you know?

Part C: Properties of Rectangles

A **rectangle** is a parallelogram that has four right angles. Find the rectangles in your set of quadrilaterals and use them to answer each question.

1. Because a rectangle is a parallelogram, what must be true about its angles, sides, and diagonals? Check your conjectures by using a protractor to measure the angles and a ruler to measure the sides.

2. Draw the diagonals of each rectangle. Use a ruler to measure their lengths. Compare their measures and write a conjecture about what you observe. Does your conjecture hold true for every rectangle? How do you know?

Part D: Properties of Squares

A **square** is a rectangle in which all the sets are the same length. Find the squares in your set of quadrilaterals and use them to answer each question.

1. Because a square is a rectangle, what must be true about its angles, sides, and diagonals? Check your conjectures by using a protractor to measure the angles and a ruler to measure the sides.

2. Because the four sides of the square are congruent, a square must also be a rhombus. Consequently, what must be true about its diagonals? Check your conjecture by using a protractor to measure the angles made by the intersecting diagonals.

Part E: Properties of Isosceles Trapezoids

A **trapezoid** is a quadrilateral with exactly one pair of parallel sides. The parallel sides are called the **bases** and the non-parallel sides are called the **legs**. An **isosceles trapezoid** is a trapezoid with two congruent legs. Find the isosceles trapezoids in your set of quadrilaterals. Use a protractor to measure its angles and a ruler to measure its sides. Record your measures.

1. Two angles that lie on the same base of an isosceles trapezoid are called **base angles**.
    a. Compare the measures of each pair of base angles in each isosceles trapezoid. Write a conjecture about what you observe. Does your conjecture hold true for all isosceles trapezoids? How do you know?
    b. Compare the measures of two angles on the different bases of each isosceles trapezoid. Write a conjecture about what you observe. Does your conjecture hold true for all isosceles trapezoids? How do you know?

2. Draw the diagonals in each isosceles trapezoid *ABCD* and then use a ruler to measure their lengths. Compare their measures and write a conjecture about what you observe. Does your conjecture hold true for every isosceles trapezoid? How do you know?

Part F: Summary and Connections

Summarize the properties of quadrilaterals by completing the following table. Place an x in the appropriate box to indicate the property holds for the given quadrilateral.

Property	Parallelogram	Rhombus	Rectangle	Square	Isosceles Trapezoid
Opposite sides parallel					
Opposite sides congruent					
All sides congruent					
Consecutive angles supplementary					
Opposite angles congruent					
All angles congruent					
Diagonals bisect opposite angles					
Diagonals perpendicular					
Diagonals bisect each other					
Diagonals congruent					

Exploration 10.9 Paper Folding and Polygons

Purpose: Students use paper folding to investigate the properties of polygons.

Materials: A compass, scissors, a protractor, and paper

Part A: Paper Folding and Polygons

Use a compass to draw a circle on a sheet of paper. Make the radius at least 3 in. Mark the center of the circle and label it *C*. Next, place a point *X* on the circumference of the circle. Use a protractor to locate points *Y* and *Z* on the circumference so that ∠*XCY* is 120° and ∠*XCZ* is 120°. Cut out the circle, but be sure you can still see points *X*, *Y*, and *Z*. Fold point *X* onto the center of the circle, *C*, and crease. Repeat with points *Y* and *Z*.

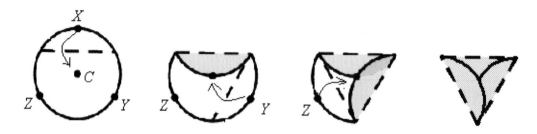

1. What polygon is created by the three folds? What are the properties of this polygon? Throughout the rest of the exploration, assume the area of the triangle is 1 unit.

2. Make a light crease to find the midpoint of one side of your triangle. Mark this point with your pencil. Fold the opposite vertex of the triangle to the midpoint and crease.
   a. What polygon have you created? What are the properties of this polygon?
   b. What is the area of this polygon if the area of the triangle is 1 unit?

3. The polygon created in the previous step consists of three congruent triangles. Fold one of the end triangles over the top of the middle triangle.
   a. What polygon have you created? What are the properties of this polygon?
   b. What is its area?

4. Fold the remaining triangle onto the other two.
   a. What polygon have you created? What are its properties?
   b. What is its area?

5. Place the three folded-over triangles in the palm of your hand. Allow the triangles to open so that a three-dimensional shape is formed.
   a. What 3-dimensional shape is made?
   b. What is its surface area?

6. Open your folded paper to get back to the large equilateral triangle you first made. Fold each vertex to the center of the circle.
   a. What polygon have you formed? What are its properties?
   b. What is its area?

7. Finally, unfold the circle and identify every polygon formed by the creases.
   a. How many triangles can you find? How many trapezoids?
   b. Can you find any parallelograms?
   c. What other polygons can you find?

Part B: Summary and Connections

1. Describe how you might use this activity to teach polygons in an elementary classroom. What aspects of this exploration might be particularly beneficial for teaching children about polygons?

2. Do an Internet search to find other paper folding activities related to polygons. What other polygons can you make by folding paper?

Exploration 10.10 Spatial Reasoning with Manipulatives

Purpose: Students use manipulatives to reason spatially in two dimensions.

Materials: Tangrams, pentominoes, squares tiles, and grid paper

Part A: Spatial Reasoning with Tangrams

A set of tangrams consists of two large triangles, a medium triangle, two small triangles, a square, and a parallelogram. The seven pieces can be used to create a variety of shapes. As you work, pay attention to the strategies you use and the observations you make.

1. Use all 7 tangram pieces to make each figure. Record your solutions by making a sketch of how you arranged the pieces.
    a. A rabbit.   b. A candle.   c. A cat.

2. Use tangrams to build each polygon. Record your solutions by making a sketch of how you arranged the pieces. Record all possible solutions you find.
    a. Build a rectangle first with 3 tangram pieces, then with 4, then 5, then 6, and finally with all 7 pieces.
    b. Build a trapezoid first with 3 tangram pieces, then with 4, then 5, then 6, and finally with all 7 pieces.
    c. Build a parallelogram that is not a rectangle first with 3 tangram pieces, then with 4, then 5, then 6, and finally with all 7 pieces.

Part B: Spatial Reasoning with Pentominoes

A pentomino is a two-dimensional shape composed of 5 squares. Any two squares that touch must do so completely along a side. Two pentominoes are considered to be the same if one can be flipped or rotated so that it can be placed exactly on top of the other.

1. Use 5 square tiles to create different pentominoes. Record your solutions on grid paper. How many pentominoes can you create?

2. A pentomino is named by the letter of the alphabet its shape most closely resembles. What letter would you assign to each pentomino? When you finish, check with your instructor or do an Internet search to find the conventional names for each shape.

3. Use Grid A and Grid B to complete the following tasks. Assume that dark regions on the grids are spaces to be left uncovered by pentominoes.
   a. Use 4 different pentominoes to fill Grid A.
   b. Use 4 different pentominoes to fill Grid B.
   c. Can you find 8 different pentominoes to fill Grids A and B at the same time?

Grid A   Grid B

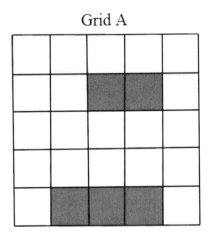

4. a. Use the pentominoes U, V, X, and Y to build two congruent figures. Each figure should contain two pentominoes. Can you create two other congruent figures, each of which is made from just two pentominoes?
   b. Use the pentominoes L, N, P, U, V, and Z to build two congruent figures. Each figure should contain three pentominoes. Can you create two other congruent figures, each of which is made from just three pentominoes?

5. Which pentominoes can be folded into an open-top box; that is, a cube with one of its faces missing? Try to use visualization skills to predict which pentominoes will make a box before you cut them from paper and fold them.

Part C: Summary and Connections

1. Take a moment to look back at what you have done.
   a. Identify any thinking strategies you used while figuring out how to create each of the shapes in Parts A and B.
   b. What observations did you make as you worked with the tangrams and pentominoes? Describe any new patterns, discoveries, conjectures, and questions that became evident to you.

2. a. How did you use spatial reasoning to complete the activities in this exploration?
   b. Do you think spatial reasoning is an important skill for elementary students to develop? Why or why not?

Exploration 10.11 Angle Properties of Polygons

Purpose: Students use measurement to discover properties of polygons.

Materials: Regular Polygon Sheet, ruler, protractor, unlined paper, and a pencil

Part A: Sum of the Interior Angles of a Polygon

Use a ruler to draw any triangle, quadrilateral, pentagon, and hexagon on several sheets of paper. Make the shapes convex and large enough so that their angles can be measured with a protractor. Measure every angle with your protractor and write the angle measure inside the angle.

1. a. Add the angle measures for each polygon and record the sum in the given blank.

   Triangle _____ Quadrilateral _____

   Pentagon _____ Hexagon _____

   b. What pattern do you see between the number of sides of the polygon and the sum of the interior angle measures? Write a formula that expresses this relationship.
   c. Use your formula to find the sum of the interior angle measures in each convex polygon.
      i. Heptagon   ii. Decagon   iii. 17-gon   iv. 50-gon

2. The sum of the interior angle measures of a convex polygon can also be found by using diagonals. A **diagonal** of a polygon is any line segment that joins two non-adjacent vertices. Draw all the diagonals from a single vertex in the quadrilateral, pentagon, and hexagon. This divides each polygon into triangles.
   a. Record the number of triangles for each polygon in the given blank.

   Quadrilateral _____ Pentagon _____ Hexagon _____

   b. What pattern do you see between the number of sides of the polygon and the number of triangles? Write a formula that expresses this relationship.
   c. Use your answer to the previous question and the fact that the sum of the interior angles in a triangle is 180° to write a formula that can be used to find the sum of the interior angles of any polygon. How does your formula compare to your formula from Question 1?

Part B: Angle Measures in Regular Polygons

1. a. In Part A, you found a formula for the sum of the interior angles of any convex polygon. In each of the regular polygons on the Regular Polygon Sheet, use a protractor to find the degree measure of an interior angle. Record the measure in the given blank.

   Triangle _____   Quadrilateral _____

   Pentagon _____   Hexagon _____

   b. What pattern do you see between the number of sides of the regular polygon and the measure of its interior angle? Write a formula that expresses this relationship.
   c. Use your formula to find the measure of an interior angle in each regular polygon.
      i. Heptagon   ii. Decagon   iii. 17-gon   iv. 50-gon

2. The **center** of a regular polygon is the center of its circumscribed circle. In regular polygons with an odd number of sides, the center is at the intersection of the segments that extend from a vertex and are perpendicular to the opposite side. In regular polygons with an even number of sides, the center is the intersection of the diagonals connecting opposite vertices. Draw the center of each regular polygon. Use a protractor to measure the **central angle**, which is the angle formed by the segments drawn from the center to two consecutive vertices.

   a. Record the measure of each central angle in the given blank.

   Triangle _____   Quadrilateral _____

   Pentagon _____   Hexagon _____

   b. What pattern do you see between the number of sides of the regular polygon and the measure of its central angle? Write a formula that expresses this relationship.
   c. Use your formula to find the measure of a central angle in each regular polygon.
      i. Heptagon   ii. Decagon   iii. 17-gon   iv. 50-gon

Part C: Summary and Connections

1. Which method do you think was the most effective way to discover the formula for the sum of the interior angle measures of a convex polygon? Why did this method make the most sense to you?

2. Can you find a formula that relates the number of sides of a convex polygon to the total number of diagonals in the polygon?

3. Can you find a formula that relates the number of sides of a regular polygon to the measure of its exterior angle?

4. Can you find a formula that relates the number of sides of a regular polygon to the number of its lines of symmetry?

Regular Polygon Sheet

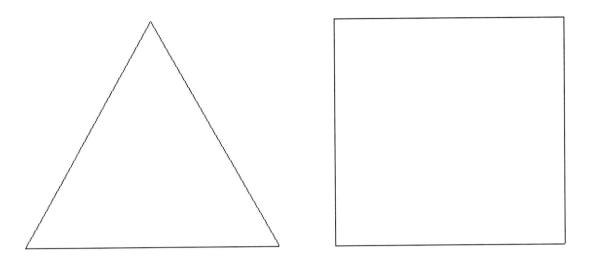

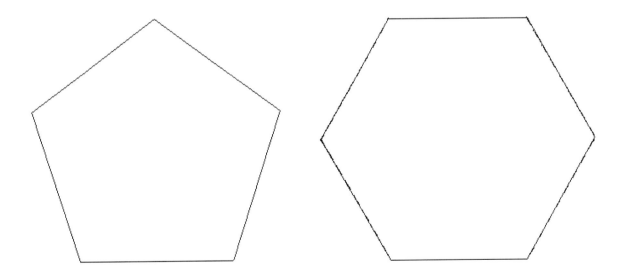

Exploration 10.12 Spatial Reasoning in Three Dimensions

Purpose: Students use spatial reasoning to make three-dimensional drawings.

Materials: 1 in. cubes, grid paper, isometric dot paper

Part A: Drawing Blocks from Different Perspectives

In the classroom, children use and create two-dimensional drawings of three-dimensional shapes. This is difficult for many students because it relies on spatial reasoning skills that are not well-developed. One way to work on these skills is to draw simple sets of blocks from different perspectives.

1. Use 1 in. cubes to create each of the following three-dimensional shapes. Next, draw the front, right, and top view of each figure on grid paper.

    a.

    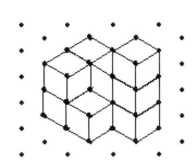

    b.

    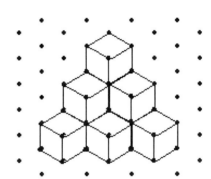

    c.

    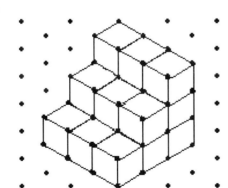

    d.

    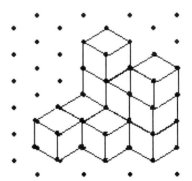

266 Chapter 10 Geometrical Shapes

2. The front, right, and top view of a three-dimensional shape are given. Use the different perspectives and 1-in. cubes to create the three-dimensional shape. Draw the shape on isometric dot paper.

   a.
   b.
   c.

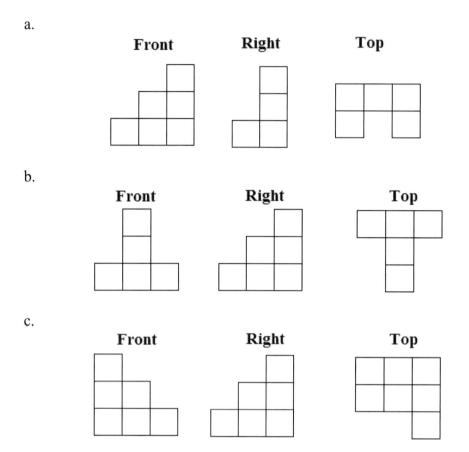

Part B: Drawing with Perspective

Drawing with perspective is a challenge for many students because they do not know which edges of a three-dimensional shape should show and which should remain hidden. Below are the outlines of several shapes. On one copy, draw the segments you would see if you were looking down at the top of the figure. On the other, draw the segments you would see if you were looking up from bottom. Leave hidden segments as dashed lines.

   1.

© 2014 Cengage Learning. All Rights Reserved. May not be scanned, copied or duplicated, or posted to a publicly accessible website, in whole or in part.

2.

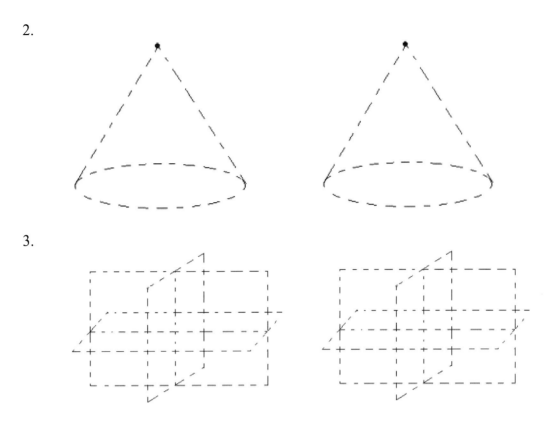

3.

Part C: Summary and Connections

1. Which part of this exploration was most difficult for you? Can you explain why?

2. Search the Internet to find other activities that can be used to improve spatial reasoning skills in three dimensions. Complete the activities and write a paragraph or two explaining which activities you think would be worthwhile to include in your classroom.

Exploration 10.13 Three-Dimensional Shapes and Their Nets

Purpose: Students use nets to investigate three-dimensional shapes.

Materials: 1-in. grid paper, scissors, and tape

Part A: Nets of a Box

A **net** is a two-dimensional representation of a three-dimensional shape that is made by cutting enough of the shape's edges so that it can be unfolded and laid flat. A three-dimensional figure can have many nets. For instance, two nets for one box are shown.

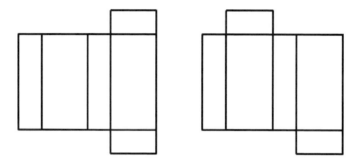

With a group of your peers, create as many non-congruent nets as you can for a box that measures 8 × 4 × 2 in. Use 1-inch grid paper to create the nets.

Part B: Nets of An Open-topped Cube

Consider an open-topped cube; that is, a cube with only five sides. Use a colored pencil to mark the edges on the cube that must be cut so that it will unfold to create the net to its right. Assume the square marked B represents the bottom face of the cube.

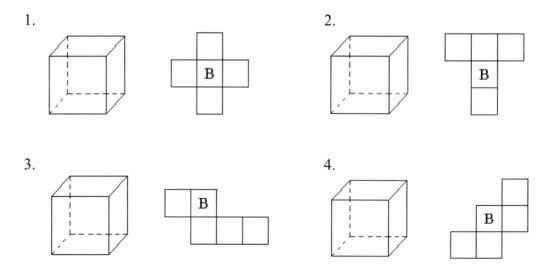

Part C: The Nets of a Pyramid

Draw a square with sides 8 in. long on 1-in. grid paper. Next, draw four congruent isosceles triangles that have sides of length 8 in., 5 in., and 5 in. The height of each triangle should be 3 in. Cut out the 5 polygons, and then fit them together to form a square pyramid. With your group, create as many non-congruent nets as possible for the square pyramid. Record your solutions by sketching the nets. Be sure to indicate the lengths of the sides in you diagram.

Part D: What 3-D Shape Does the Net Make?

Several nets of three-dimensional shapes are shown. Predict the three-dimensional shape the net will create by mentally folding it together. After you predict the three-dimensional shape, check your prediction by tracing the net, cutting it out, and folding and taping it together.

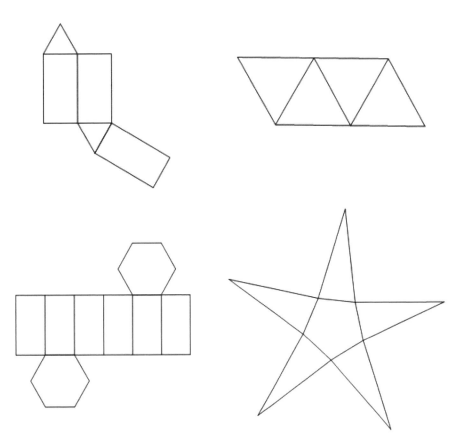

Part E: Summary and Connections

1. Did you find it easier to go from the net to the three-dimensional shape or from the three-dimensional shape to the net? Why do you think this was so?

2. What is the connection between the surface area of a three-dimensional shape and the area of its net?

Exploration 10.14 Regular and Semi-Regular Polyhedra

Purpose: Students use models to investigate regular and semi-regular polyhedra.

Materials: A set of paper or plastic congruent equilateral triangles, congruent squares, congruent regular pentagons, and congruent hexagons

Part A: Regular Polyhedra

Form a small group with two or three of your peers to complete the following tasks involving regular polyhedra. A **regular polyhedron** is a convex polyhedron that has congruent dihedral angles and faces that are the same congruent regular polygon.

1. Use congruent equilateral triangles to create as many regular polyhedra as possible.
    a. How many regular polyhedra were you able to create?
    b. Why do you think you have created all that you can?

2. Use congruent squares to create as many regular polyhedra as possible.
    a. How many regular polyhedra were you able to create?
    b. Why do you think you have created all that you can?

3. Use congruent regular pentagons to create as many regular polyhedra as possible.
    a. How many regular polyhedra were you able to create?
    b. Why do you think you have created all that you can?

4. Use congruent regular hexagons to create as many regular polyhedra as possible.
    a. How many regular polyhedra were you able to create?
    b. Why do you think you have created all that you can?

5. a. Do you think it is possible to create a regular polyhedron with congruent regular heptagons? Octagons? Explain.
   b. How many different congruent regular polygons can be used to make any one regular polyhedron?

6. Summarize your findings by describing each regular polyhedron. Give the number and the shape of the faces, the number of vertices, and the number of edges. Find the name of each of these regular polyhedra.

Part B: Semi-Regular Polyhedra

A **semi-regular polyhedron** is a convex polyhedron made from congruent regular polygons, but the faces are not limited to a single regular polygon. For instance, a semi-regular polyhedron can be composed of congruent equilateral triangles and congruent squares, provided the configuration of polygons is the same at each vertex.

1. What is the difference between a regular polyhedron and a semi-regular polyhedron?

2. Thirteen semi-regular polyhedra exist. Use your materials to create one with equilateral triangles and squares.

3. No regular polyhedron uses a regular hexagon as a face. However, five semi-regular polyhedra do. Create at least two.

4. What is the fewest number of faces that a semi-regular polyhedron can have? Explain.

5. Susie claims that no semi-regular polyhedron can be created with a regular polygon having more than six sides. Do you agree or disagree? Explain.

Part C: Summary and Connections

1. Use toothpicks and mini-marshmallows to create the regular polyhedra. What, if any, are the advantages of these models over the ones you created in Part A?

2. Draw a net for each regular polyhedron.

Exploration 10.15 Euler's Formula

Purpose: Students construct three-dimensional figures to discover Euler's formula.

Materials: Polydrons® or other materials to build polyhedrons

Part A: Discovering Euler's Formula

The number of faces, vertices, and edges of a polyhedron are not independent of one another. Use your materials to construct the following polyhedra:

Triangular prism	Rectangular prism	Hexagonal prism
Triangular pyramid	Square pyramid	Pentagonal pyramid
Octahedron	Cube octahedron	Truncated tetrahedron

After constructing each polyhedron, count the faces, vertices, and edges. Record the information in the following table.

Polyhedron	Number of Faces	Number of Vertices	Number of Edges
Triangular prism			
Rectangular prism			
Hexagonal prism			
Triangular pyramid			
Square pyramid			
Pentagonal pyramid			
Octahedron			
Cube octahedron			
Truncated tetrahedron			

Let f, v, and e represent the number of faces, vertices, and edges of a polyhedron. Euler's formula states a relationship among the three variables. Use the numbers in the table to find the relationship, and then state it as a formula in terms of f, v, and e.

Part B: Using Euler's Formula

Use Euler's formula to answer each question.

1. a. A polyhedron has 11 vertices and 20 edges. How many faces does it have?
   b. A polyhedron has 26 vertices and 15 faces. How many edges does it have?
   c. A polyhedron has 12 faces and 30 edges. How many vertices does it have?

2. Show that Euler's formula holds for a heptagonal prism.

3. Show that Euler's formula holds for an icosahedron.

Part C: Summary and Connections

1. What is the fewest number of faces a polyhedron can have? Vertices? Edges?

2. Is it possible for a polyhedron to have 14 faces and 16 edges? Why or why not?

Chapter 11
Congruence, Similarity, and Constructions

In the last chapter, we analyzed two- and three-dimensional shapes for their defining characteristics and properties. We now continue our study of shapes by considering congruence, similarity, and geometric constructions.

Congruence and similarity are two important relationships that exist between geometric figures. **Congruence** is the notion that two figures have the same shape and size, whereas **similarity** is the notion that two figures have the same shape, but not necessarily the same size. Both relationships can be applied to a variety of shapes and can be used as powerful tools for solving problems. In the elementary classroom, children come to understand congruence and similarity by manipulating physical shapes. To determine whether two shapes are congruent, they learn to place one over top of the other to see if they match. For similarity, they compare figures of different sizes to see if they have same basic shape. As students progress into the middle grades, they approach congruence and similarity more formally by using reasoning and measurement to compare corresponding parts of figures.

The second topic in this chapter is geometric constructions. The very nature of geometry makes it necessary to use pictures to clarify concepts, solve problems, and verify properties. Some of these situations require more precision, which can be obtained by using constructions. In the classroom, constructions are often delayed until the middle grades when students have the physical coordination to use a compass and straightedge. However, there are a number of tools and strategies, such as plastic reflectors and paper folding, which make constructions accessible to students at an earlier age.

The explorations in Chapter 11 focus on congruence, similarity, and geometric constructions. Specifically,

- Explorations 11.1 - 11.2 consider congruent triangles and other shapes.
- Explorations 11.3 - 11.5 focus on similar shapes and self-similarity.
- Explorations 11.6 - 11.12 use a variety of tools and computer software to complete geometrical constructions.

Exploration 11.1 Congruent Triangles and Conditions for Congruency

Purpose: Students develop triangle congruence theorems by building triangles.

Materials: Fettuccine, markers, modeling clay, tape, protractor, and paper

Part A: Building Triangles

Cut the fettuccine into lengths of 3 in., 4 in., 5 in., 6 in., and 7 in. Color the fettuccine so that all the 3 in. lengths are one color, all the 4 in. lengths are another color, and so on. Use the fettuccine, modeling clay, and protractor to build all possible non-congruent triangles for each of the given specifications. Tape the triangles to paper, and record any observations you make as you build the triangles.

1. Triangles with sides of 3 in., 4 in., and 6 in.

2. Triangles with sides of 5 in. and 7 in. and an included angle of 50°.

3. Triangles with sides of 5 in. and 7 in. and an angle of 50° so that the 5-in. side is opposite the 50° angle.

4. Triangles with angles of 35° and 60° and an included side of 7 in.

5. Triangles with angles of 35° and 60° and a side of 7 in. so that the 35° angle is at one end of the 7 in. side but the 60° angle is not at the other.

6. Triangles with angles of 20°, 40°, and 120°.

Part B: Interpreting the Results

1. a. Which sets of specifications in Part A led to unique triangles? Can you explain why?
   b. For the specifications that led to a unique triangle, change the lengths of the sides and the sizes of the angles. Do you still obtain a unique triangle? Build several new triangles to test your conjecture.
   c. What is the significance of getting a unique triangle from a given set of specifications?

2. Intuitively, two triangles are **congruent** if they have the same shape and size.
   a. If a set of specifications leads to a unique triangle and two triangles satisfy those specifications, what is true about the triangles? Can you explain why?
   b. Generalize your observations from Question 1. That is, for each set of specifications, write a summary sentence that states the relationship between two triangles that satisfy those specifications. For instance, for the first set of

specifications you might write, "If three sides of one triangle are congruent respectively to three sides of another triangle, then the triangles are congruent."

3. More formally, two triangles are congruent if there is a correspondence between their vertices such that each pair of corresponding segments and each pair of corresponding angles have the same measure. Based on your answers to the previous questions, why is it not necessary to show that all six pairs of corresponding parts are congruent in order to ensure that two triangles are congruent?

Part C: Summary and Connections

Repeat the exploration with the following specifications for right triangles. Determine which specifications lead to unique right triangles and then generalize your observations. If necessary, cut new lengths of fettuccine to complete the triangles.

1. Right triangles with legs of length 3 in. and 4 in.

2. Right triangles with a hypotenuse of 5 in.

3. Right triangles with a hypotenuse of 5 in. and one acute angle of 40°.

4. Right triangles with acute angles of 30° and 60°.

Exploration 11.2 Congruent Shapes

Purpose: Students use manipulatives to explore congruent shapes.

Materials: 5 × 5 geoboard, rubber bands, dot paper, tangrams

Part A: Congruent Shapes on a Geoboard

In general, two shapes are **congruent** if they have the same shape and size. One way to better understand congruence is to analyze shapes using a variety of manipulatives. For instance, consider the following activities for a 5 × 5 geoboard.

1. Use rubber bands to create the following triangles on a geoboard. Which triangles are congruent? What process did you use to make your determination?

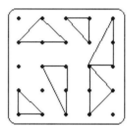

2. a. The following diagrams illustrate different ways to divide the 4 by 4 square on the geoboard into two shapes. Are the shapes congruent? How do you know?

   i. 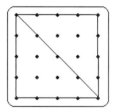   ii. 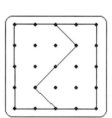   iii.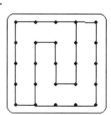

   b. How many other ways can you divide the 4 by 4 square into 2 congruent shapes? Record your answers on grid paper.

   c. Which of the following shapes can be divided into two congruent shapes? For those that can, show how.

   i. 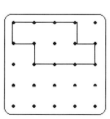   ii. 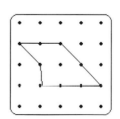   iii.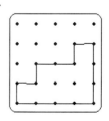

280   Chapter 11 Congruence, Similarity, and Constructions

3. The following shape can be divided into 2, 3, 4, 6, 8, 12, 24, and 48 congruent shapes. Determine how to do so and then record your answers on grid paper.

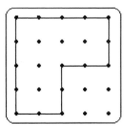

Part B: Congruent Shapes with Tangrams

1. How many different congruent trapezoids can you make using any two tangram pieces? What about parallelograms? Other shapes? In each case, how do you know the shapes are congruent?

2. How many different congruent trapezoids can you make using any three tangram pieces? What about parallelograms? Other shapes? In each case, how do you know the shapes are congruent?

3. How many different ways can you use tangram pieces to make a triangle that is congruent to one of the large triangles? Record your answers on paper.

Part C: Formalizing the Idea of Congruence

One way to formalize the idea of congruence is to use the measures of segments and angles. Specifically, two segments are congruent if they have the same length, and two angles are congruent if they have the same angle measure. By using congruence for these shapes, we can extend congruence to other planar shapes.

1. If the following two polygons are congruent, what must be true about their sides? What must be true about their angles?

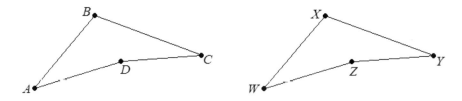

2. When determining whether two polygons are congruent, we must decide which segments and which angles to compare. How do we know which comparisons to make?

3. Use your answers to Questions 1 and 2 to write a definition of two congruent
   a. triangles.   b. quadrilaterals.   c. polygons.

4. The two figures in Question 1 have the same orientation. Must two shapes have the same orientation to be congruent?

Part D: Summary and Connections

1. Intuitively, two shapes are congruent if one shape can be placed exactly over the top of the other.
    a. How does this idea of congruence match up with the formal statements you made about congruence in Part C?
    b. What other geometrical ideas do you use when thinking of congruence in this way?

2. Use what you know about each figure to answer each question.
    a. If two corresponding sides of two squares are congruent, are the squares congruent?
    b. If two corresponding sides of two rectangles are congruent, are the rectangles congruent?
    c. If two pairs of corresponding sides of two parallelograms are congruent, are the parallelograms congruent?
    d. If two corresponding sides of two regular hexagons are congruent, are the hexagons congruent?
    e. If the radii of two circles are congruent, are the circles congruent?

Exploration 11.3 Measuring with Similar Triangles

Purpose: Students use similar triangles to measure distances and heights.

Materials: A tape measure or yard stick, two strings at least 20 ft long, and tape

Part A: How High Can You Reach?

Form a small group with two or three of your peers. Tape one end of a piece of string to the floor. Have one person hold the other end of the string in one hand and take five or six steps from where the string touches the floor, reaching as high as possible with the string so that it becomes tight. Next, have another person pick a spot close to the place where the string is taped and use the measuring stick to measure the distance from the floor to the string at the spot. This forms two similar triangles as shown in Figure 1.

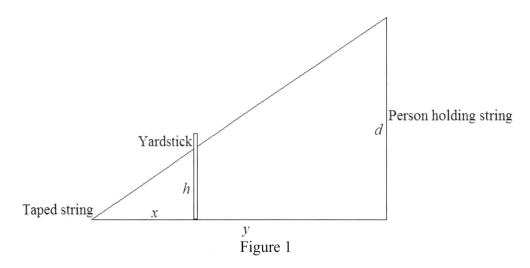

Figure 1

Record the height (h) of the string at this point. Next, measure the distance from the point on the floor to the yardstick (x), the distance from the point on the floor to the person holding the string (y), and the distance from the floor to the point at which the person is holding the string (d). Record the measures.

$d =$ _____ $h =$ _____ $x =$ _____ $y =$ _____

1. Why are the two triangles in Figure 1 similar?

2. Use proportional reasoning and the values of h, x, and y, to compute an estimate for d.
   a. How does your computed estimate for d compare to your measured value for d?
   b. What might account for any differences between the values?

Part B: Measuring the Distance Between Two points on the Floor

Take a piece of sting 10 ft long, stretch it out taut, and tape both ends to the floor. From one end, mark a length of 2 ft along the string. Take a second piece of string 10 ft long and mark a length of 2 ft from one of its ends. Stretch the second string taut so that both strings cross one another at the 2 ft marks. (See Figure 2). Tape the ends to the floor.

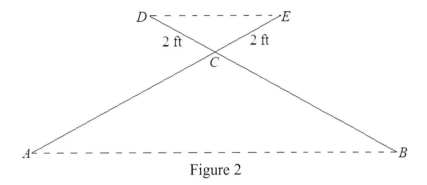

Figure 2

Measures the lengths *DE*, *BC*, and *AB*. Record the measures.

DE = _____ BC = _____ AB = _____

1. Why are the two triangles in Figure 2 similar?

2. Use proportional reasoning to estimate the length of $\overline{AB}$.
   a. How does your computed estimate for *AB* compare to your measured value of *AB*?
   b. What might account for any differences between the values?

Part C: Using Your Own Method

Devise a strategy for using similar triangles to measure a distance or length that is different from the two used in this exploration. Write a paragraph or two explaining what you measured, the method you used, how you knew the triangles were similar, and how your computed estimate compared to the measured distance or length.

Part D: Summary and Connections

1. What is the value in using similar triangles to measure? Do you think similar triangles provide an accurate way to measure? Why or why not?

2. How might you adapt this exploration for use with a group of middle grades students?

Exploration 11.4 Similar Shapes

Purpose: Students use measurement to explore similar shapes.

Materials: Colored paper (red, blue, and green), rulers, protractors, and scissors

Part A: Similar Triangles

Form a small group with two of your peers. Give each person three sheets of colored paper so that each person has only one color. Using a protractor and a ruler, have each person construct three triangles, one per sheet of paper. The sides of the triangles can be any length as long as the triangles meet the following specifications:

Triangle #1 has angles measuring 30° and 70°
Triangle #2 has angles measuring 40° and 90°
Triangle #3 has angles measuring 35° and 110°

After making the triangles, each person should number their triangles 1, 2, and 3 and then measure the length of the sides in each triangle to the nearest tenth of a centimeter. Record the measures on the inside of the triangle next to the appropriate side. Next, each person should measure the angles and record the angle measures on the inside of each triangle. Finally, cut out the triangles. When finished, the group should have nine triangles: three of each color and three with each set of specifications.

1. Visually compare the three triangles labeled number 1, the three triangles labeled number 2, and the three triangles labeled number 3. What do you observe? What conjectures can you make? Record your group's observations and conjectures.

2. Complete the following tables by taking the ratios of the lengths of the indicated sides in each set of triangles.

Triangle #1

Triangles to compare	$\dfrac{\text{long side}}{\text{long side}}$	$\dfrac{\text{middle side}}{\text{middle side}}$	$\dfrac{\text{short side}}{\text{short side}}$
Red to blue			
Red to yellow			
Blue to yellow			

Triangle #2

Triangles to compare	$\dfrac{\text{long side}}{\text{long side}}$	$\dfrac{\text{middle side}}{\text{middle side}}$	$\dfrac{\text{short side}}{\text{short side}}$
Red to blue			
Red to yellow			
Blue to yellow			

Triangle #3

Triangles to compare	$\dfrac{\text{long side}}{\text{long side}}$	$\dfrac{\text{middle side}}{\text{middle side}}$	$\dfrac{\text{short side}}{\text{short side}}$
Red to blue			
Red to yellow			
Blue to yellow			

3. Use your observations from the three tables to answer each question.
    a. What can you conclude about the sides of any two triangles that have the same angle measures?
    b. What can you conclude about any two triangles that have the same angle measures?

Part B: Similar or Not Similar?

Similarity applies not only to triangles, but to many other shapes as well.

1. Using what you have observed about similar triangles, what must be true about any two shapes for them to be similar?

2. Use your notion of similarity to determine whether each pair of shapes is similar. In each case, explain your reasoning.

   a.

   b.

   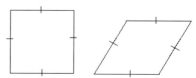

   c.

   d.

   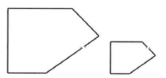

3. Use your notion of similarity to answer each question. If the shapes are similar, explain why. If not, give a counterexample.
    a. Is a rectangle with sides 5 cm and 7 cm similar to a rectangle with sides of 10 cm and 13 cm?
    b. Are any two squares similar? Any two rectangles? Any two parallelograms?
    c. Are any two regular polygons with the same number of sides similar?
    d. Are any two circles similar?

Part C: Summary and Connections

1. If two shapes are similar, are they congruent? If two shapes are congruent, are they similar? Explain your thinking.

2. Similarity is often introduced in the elementary grades. Examine several sets of elementary curriculum materials to see how similarity is taught.
    a. How does your notion of similarity compare to those presented in the curriculum materials?
    b. If similar shapes are presented in different ways, which way do you think is most effective?

Exploration 11.5 Rep-tiles and Fractals

Purpose: Students use rep-tiles and fractals to explore self-similarity.

Materials: Patterns blocks, ruler, paper, scissors, and Internet access

Part A: Rep-tiles

A shape is self-similar if it looks "roughly" the same when any part of the figure is magnified or reduced. For instance, a **rep-tile** is a self-similar figure that is made by dissecting a polygon into smaller versions of itself or by fitting copies of the polygon together to form a larger, similar version of itself. Figure 1 illustrates an example of a triangular rep-tile.

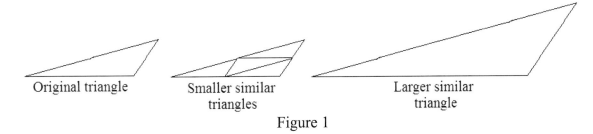

Original triangle   Smaller similar triangles   Larger similar triangle

Figure 1

1. Use a ruler to construct a triangle with sides that are between 1 in. and 2 in. long. Cut out your triangle.
   a. Use your triangle to make a rep-tile of larger similar triangles by tracing copies of your triangle on a sheet of paper. Describe how you had to manipulate the original triangle to make your rep-tile.
   b. Trace several copies of your original triangle on another sheet of paper. Devise a strategy for making a rep-tile that consists of smaller triangles similar to the original.
   c. Can any triangle be used to create a rep-tile? Explain your thinking.

2. Use a ruler to construct a quadrilateral with sides that are between 1 in. and 2 in. long. Cut out your quadrilateral.
   a. Use your quadrilateral to make a rep-tile of larger similar quadrilaterals by tracing copies of your quadrilateral on a sheet of paper. Describe how you had to manipulate the original quadrilateral to make your rep-tile.
   b. Trace several copies of your original quadrilateral on another sheet of paper. Devise a strategy for making rep-tile that consists of smaller quadrilaterals that are similar to the original.
   c. Can any quadrilateral be used to create a rep-tile? Explain your thinking.

3. A set of pattern blocks has six basic shapes: a hexagon, a trapezoid, two rhombi, a square, and an equilateral triangle. Which of the patterns blocks can be used to make a reptile?

4. Which of the following shapes can be used to make a rep-tile?

Part B: Fractals

Another shape with self-similarity is a **fractal**, which is a geometric pattern generated by a recursively-defined function.

1. One way to draw a simple fractal is to begin with a line segment in the first iteration. In the second iteration, draw two sides of an equilateral triangle so that they divide the original segment into thirds. All subsequent iterations are drawn by placing two sides of an equilateral triangle on each segment of the previous iteration, again dividing each segment into thirds. The first three iterations are shown. Draw the fourth and fifth iteration of the fractal.

2. With the use of computers, we can visualize and explore the self-similarity of fractals quickly and efficiently. Search the Internet for a **Mandelbrot set** generator. This fractal is name after Benoit Mandelbrot who first introduced fractals in the early 1980s. Use the fractal generator to magnify the fractal a number of times.
    a. Write a paragraph or two to describe the self-similarity you observe.
    b. Once you are finished exploring the Mandelbrot set with the fractal generator, do another Internet search to find applications of the Mandelbrot set.

Part C: Summary and Connections

1. Write a paragraph or two explaining how you might use rep-tiles to help elementary children understand the idea of similarity.

2. Search the Internet for fractal art. Select several pictures and describe how you see the Mandelbrot set being used in each picture.

Exploration 11.6 Constructing Shapes With a Plastic Reflector

Purpose: Students use a plastic reflector to make constructions.

Materials: Plastic reflector, ruler, protractor, and unlined paper

Part A: Basic Constructions

One way to make constructions is to use a plastic reflector. A plastic reflector not only reflects the image of an object, but it also allows us to see through to the other side. Use a plastic reflector to complete each construction. In each case, describe the process you used to complete the construction.

1. Draw a segment on a sheet of paper. Use a plastic reflector to copy the segment onto another part of the paper. Measure the length of the original segment and the copy of the segment. How do the lengths compare?

2. Draw an angle on a sheet of paper. Use a plastic reflector to copy the angle onto another part of the paper. Measure the original angle and the copy of the angle with a protractor. How do the angles compare?

3. Draw an angle and then bisect it using the plastic reflector. Use a protractor to measure the two angles created by bisecting the original angle. How do the angles compare?

4. Draw a line l and a point P that is not on the line. Use the plastic reflector to construct a line through P that is perpendicular to line l.
   a. Use your protractor to measure the angles created by the intersecting lines. Are the angles right angles?
   b. Use your method for constructing a perpendicular line to construct the perpendicular bisector of a line segment.

Part B: Other Constructions

Use a plastic reflector to complete each construction. In each case, describe the method you used to complete the construction.

1. Draw a line l and a point P that is not on the line. Construct a line through P that is parallel to l.

2. Draw a line segment $\overline{AB}$. Construct an isosceles triangle with legs congruent to $\overline{AB}$.

3. Draw a line segment $\overline{AB}$. Construct an equilateral triangle with sides congruent to $\overline{AB}$.

4. Draw a line segment $\overline{AB}$. Construct a rhombus with sides congruent to $\overline{AB}$.

5. Draw a line segment $\overline{AB}$. Construct a square with sides congruent to $\overline{AB}$.

6. Which shapes were the most difficult to construct with a plastic reflector? What aspects of the constructions made them difficult to complete?

Part C: Summary and Connections

1. Do you think the constructions that you made with a plastic reflector were accurate? That is, did you actually construct the shape you were asked to construct?

2. Why might a plastic reflector be a better choice than a straightedge and compass for making constructions in the elementary and middle grades?

Exploration 11.7 Basic Constructions with Geometry Software

Purpose: Students become familiar with the basic features of geometry software by completing simple constructions.

Materials: Geometry software

Part A: Points, Segments, Rays, and Lines

Geometry software takes the form of an electronic drawing pad that allows the user to make geometrical constructions. In this exploration, you will use some of the basic features of these programs to construct basic shapes. Because most geometrical shapes are made from points and segments, we begin with these shapes.

1. One way to create segments, rays, and lines is to use a line tool. Use the line tool to create several segments, rays, and lines. Describe how you used the line tool to do so.
    a. How many points are highlighted on each segment, ray, or line that you draw? Why does it make sense for the program to highlight this number of points?
    b. What happens when you use the selection tool to click and drag one of the points on a segment, ray, or line?

2. Another way to create segments, rays, and lines is to construct them using two points. Use the point tool to place several points anywhere on the screen. Highlight points and construct segments, rays, or lines between them.
    a. How many points must be highlighted before the software will allow you to construct a segment, ray or line? Why does this make sense?
    b. When constructing a ray, how does the software determine the direction of the ray?
    c. What happens if you select three or four points at once and then construct segments between them?

3. Use the text tool to label some of the points, segments, rays, and lines. Describe the process you used to do so.

Part B: Angles

Geometry software can also be used to perform constructions with angles. Clear your screen from the previous exploration.

1. Place three points on the screen and label them *A*, *B*, and *C*. Create ∠*ABC* without creating a triangle between the three points. Describe the method you used to do so.

2. Use the software to construct the angle bisector of ∠ABC. Describe the process you used to do so.
   a. Click and drag point A. What happens to the angle bisector as you move A?
   b. What happens if you select A, B, and C in a different order when constructing and angle bisector?

Part C: Circles

Geometry software can also be used to create circles in two ways. Clear your screen from the previous constructions.

1. One way to create a circle is to use the compass tool. Use the compass tool to create a circle.
   a. After creating a circle, how many points does the software highlight? Why does this make sense?
   b. What happens when you click and drag the center of the circle? What about other points on the circle?

2. Another way to create a circle is to construct it from a point and a segment. Use the segment tool to construct a segment. Label its end points A and B. Next, use the point tool to place a point anywhere on the screen. Label it C. Use the software to construct a circle with center C and a radius congruent to $\overline{AB}$.
   a. Describe the process you used to complete the construction.
   b. Click and drag A. What happens to the circle?
   c. Click and drag C. What happens to the circle?

Part D: Summary and Connections

1. Continue to explore the software. What constructions will the software complete with a simple command? What basic shapes must be highlighted to complete each construction?

2. How does this activity illustrate the difference between a drawing and a construction?

3. What are the advantages and disadvantages of using geometry software to make constructions?

Exploration 11.8 Constructing Parallel and Perpendicular Lines

Purpose: Students complete constructions with parallel and perpendicular lines using geometry software.

Materials: Geometry software

Part A: Constructions with Perpendicular Lines

Geometry software can be used to complete a number of constructions with perpendicular and parallel lines. We begin with perpendicular lines.

1. Place two points anywhere on the screen and label them *A* and *B*. Construct a line between them. Next, construct a point on the line and label it *C*. Use the software to construct a line perpendicular to $\overleftrightarrow{AB}$ that passes through *C*.
   a. What happens to the lines if you drag either *A* or *B*?
   b. What happens to the lines if you drag *C*?

2. Place another point anywhere on the screen that is not on $\overleftrightarrow{AB}$. Label the point *D*. Use the software to construct a line perpendicular to $\overleftrightarrow{AB}$ that passes through *D*.
   a. What happens to the lines if you drag either *A* or *B*?
   b. What happens to the lines if you drag *C*?
   c. What must be true about the line that goes through *C* in relationship to the line that goes through *D*?

3. Once you are finished exploring these lines, clear your screen and construct a segment $\overline{EF}$. Use the software to construct the perpendicular bisector of $\overline{EF}$.
   a. Write a short summary of the steps you took to complete the construction.
   b. What happens to the perpendicular bisector if you drag either *E* or *F*?

4. Use a more traditional method to construct two perpendicular lines using the software. (Hint: See Example 11.22 in Section 11.3 of the text.) Write a summary of the steps you took to complete the construction.

Part B: Constructions with Parallel Lines

1. Place two points on the screen and label them *A* and *B*. Construct a line between them. Next, construct a point anywhere on the screen that is not on $\overleftrightarrow{AB}$ and label the point *C*. Use the software to construct a line parallel to $\overleftrightarrow{AB}$ that passes through *C*.
   a. What happens to the lines if you drag either *A* or *B*?
   b. What happens to the lines if you drag *C*?

2. More traditional methods can also be used to construct parallel lines with geometry software. For instance, a rhombus can be used to construct a line parallel to another line through a point not on the line. Devise a method for doing so. (Hint: See Example 11.21 in Section 11.3 of the text.) Write a summary of the steps you took to complete the construction using the software.

3. Use perpendicular lines to construct two parallel lines. Write a summary of the steps you took to complete the construction using the software.

4. Use congruent alternate interior angles to construct two parallel lines. Write a summary of the steps you took to complete the construction using the software.

Part C: Summary and Connections

1. What other methods can you use to construct perpendicular and parallel lines using geometry software?

2. What kinds of shapes require parallel and perpendicular lines to be constructed?

Exploration 11.9 Centers of Triangles

Purpose: Students use constructions to explore the four centers of triangles.

Materials: Compass, straightedge, plastic reflector, paper, scissors, and geometry software

Part A: The Centroid of a Triangle

Triangles have a number of interesting points that are often called centers of triangles. For instance, one center is found by constructing the medians of a triangle. A **median** is a segment that connects a vertex to the midpoint of the opposite side. Use your straightedge to draw a triangle on a sheet of paper. Make each side at least 4 in. long.

1. Use a straightedge and compass, a plastic reflector, or geometry software to construct the three medians of the triangle.
    a. Describe the process you used to construct the medians.
    b. What is true about the medians? Is this true for every triangle? Test your conjecture with other triangles.

2. The medians of a triangle intersect at a point called the **centroid.** Label the centroid on your paper triangle and cut out the triangle. Place a pencil upside down and then place the centroid of the triangle on the tip of the pencil. What do you notice? What does this imply about the centroid of a triangle?

Part B: The Incenter of a Triangle

Use your straightedge to draw another triangle on a sheet of paper. Again, make each side at least 4 in. long.

1. Use a straightedge and compass, a plastic reflector, or geometry software to construct the angle bisectors of the triangle.
    a. Describe the process you used to construct the angle bisectors.
    b. What is true about the angle bisectors? Is this true for every triangle? Test your conjecture with other triangles.

2. The angle bisectors of a triangle intersect at a point called the **incenter**. It is the center of the triangle's **incircle**, which is the largest circle that fits in the triangle so that the triangle's sides are tangent to the circle. Label the incenter on your paper triangle.
    a. Will the incenter of a triangle always be inside the triangle? Explain.
    b. Find the radius and draw the incircle. Describe the process you used to construct the incircle.

Part C: The Circumcenter of a Triangle

Use your straightedge to draw a third triangle with sides at least 4 in. long.

1. Use a straightedge and compass, a plastic reflector, or geometry software to construct the perpendicular bisectors of the three sides.
    a. Describe the process you used to construct the perpendicular bisectors.
    b. What is true about the perpendicular bisectors? Is this true for every triangle? Test your conjecture with other triangles.

2. The perpendicular bisectors of the sides of a triangle intersect at a point called the **circumcenter.** It is the center of the triangle's **circumcircle**, which is the circle that intersects the triangle at its vertices. Label the circumcenter on your paper triangle.
    a. Will the circumcenter of a triangle always be inside the triangle? Explain.
    b. Find the radius and draw the circumcircle. Describe the process you used to construct the circumcircle.
    c. What is special about the circumcenter in relation to the vertices of the triangle?

Part D: The Orthocenter of a Triangle

Use your straightedge to draw a fourth triangle with sides at least 4 in. long.

1. Use a straightedge and compass, a plastic reflector, or geometry software to construct the altitudes of the triangle. An **altitude** is a segment that extends from a vertex and is perpendicular to the line containing the opposite side.
    a. Describe the process you used to construct the altitudes.
    b. What is true about the altitudes? Is this true for every triangle? Test your conjecture with other triangles.

2. The three altitudes intersect at a point called the **orthocenter.** Will an altitude of a triangle always be inside the triangle? What about the orthocenter? Explain.

3. Construct a triangle using geometry software. Construct the orthocenter, centroid, and circumcenter of the triangle. Label the points, O, C, and R respectively.
    a. Construct the segment $\overline{OR}$. What do you notice about the points O, C, and R? Use the software to test your conjecture.
    b. Measure the distance from O to C and then from C to R. What do you notice about CR compared to OC? Use the software to test your conjecture.

Part E: Summary and Connections

1. Why is geometry software a valuable tool for completing this exploration?

2. In this exploration, you found the circumcenter and circumcircle of a triangle. The circumcircle is said to circumscribe the triangle. What other shapes can be circumscribed and how would you construct the circumcircle?

Exploration 11.10 Paper Folding and Constructions

Purpose: Students perform constructions by folding paper.

Materials: Patty paper, ruler, and compass

Part A: Constructing the Perpendicular Bisector of a Segment

1. Use a ruler to draw a segment $\overline{AB}$ on patty paper.
    a. Visualize where the perpendicular bisector of $\overline{AB}$ is located. How can you fold the paper so that the crease is the perpendicular bisector of $\overline{AB}$?
    b. Once you decide where to fold, make the fold and crease. How do you know that the crease is the perpendicular bisector of the segment?

2. Choose a point C on the perpendicular bisector and measure the distances from C to A and from C to B. What do you observe?

3. Choose another point D on the perpendicular bisector and measure the distances from D to A and D to B. What do you observe?

4. Based on your observations in the previous two questions, what conjecture can you make about any point on the perpendicular bisector of a segment? Can you prove it?

Part B: Constructing the Bisector of an Angle

1. Use a ruler to draw an angle on a piece of patty paper.
    a. Visualize where the angle bisector is located. How can you fold the paper so that the crease is the bisector of the angle?
    b. Once you decide where to fold, make the fold and crease. How do you know that the crease is the bisector of the angle?

2. Choose a point C on the angle bisector. Measure the distance from C to each side of the angle. What do you observe? (Remember the distance from a point to a line is the length of the perpendicular segment from the point to the line.)

3. Choose another point D on the angle bisector. Measure the distance from D to each side of the angle. What do you observe?

4. Based on your observations in previous two questions, what conjecture can you make about the points on an angle bisector?

Part C: Other Constructions

Use patty paper to complete the following construction. In each case, explain why your construction works.

1. Draw a segment $\overline{AB}$ on patty paper. Construct an isosceles triangle with legs congruent to $\overline{AB}$.

2. Draw a segment $\overline{AB}$ on patty paper. Construct an equilateral triangle with sides congruent to $\overline{AB}$.

3. Draw a segment $\overline{AB}$ on patty paper. Construct a square with sides congruent to $\overline{AB}$.

4. Use a compass to draw a circle on patty paper. Construct a regular octagon by folding the paper to locate the vertices.

5. What other shapes can you construct with patty paper?

Part D: Summary and Connections

1. How are constructions with patty paper similar to or different from constructions with a plastic reflector? What about a straightedge and compass?

2. What advantages and disadvantages might there be in using patty paper to make constructions in the classroom?

Exploration 11.11 Constructing Triangles and Quadrilaterals

Purpose: Students construct quadrilaterals using geometry software.

Materials: Geometry software

Part A: Constructing Triangles

Geometry software can be used to complete a number of constructions with polygons. We begin with triangles. Clear your screen after each construction.

1. Use the segment tool to create a segment anywhere on the screen. Label the endpoints A and B. Construct two circles using points A and B as the centers and segment $\overline{AB}$ as the radius. Construct one of the intersection points of the resulting circles and then hide the circles. Label the new point C and construct $\triangle ABC$.
    a. What kind of triangle is $\triangle ABC$? How do you know?
    b. What happens if you click and drag A or B? What about C?

2. Construct an isosceles triangle that is not equilateral. Be sure that the triangle remains isosceles as you click and drag its vertices. Write a short summary of the steps you took to complete the construction.

3. Construct a right triangle. Be sure that the triangle remains right as you click and drag its vertices. Write a short summary of the steps you took to complete the construction.

4. Using the right triangle from the previous construction, highlight the sides and measure their lengths. Next, use the software to compute the square of each length and then add the squares of the two legs.
    a. What do you notice about the sum of the squares of the legs compared to the square of the hypotenuse?
    b. Click and drag one of the vertices. Does the relationship between the sides always hold true?

Part B: Constructing Quadrilaterals

Use geometry software to complete each construction involving a quadrilateral. Clear the screen after each construction.

1. Construct a parallelogram. Be sure the figure remains a parallelogram as you drag any of its vertices.
    a. Write a short summary of the steps you took to complete the construction.
    b. Use the software to measure the angles. What remains true about the opposite angles as you drag a vertex? What remains true about the consecutive angles?

c. Use the software to measure the sides. What remains true about the opposite sides as you drag a vertex?

d. What do your answers to the previous questions imply about the angles and sides of any parallelogram?

2. Construct a rhombus. Be sure the figure remains a rhombus as you drag any of its vertices.

   a. Write a short summary of the steps you took to complete the construction.

   b. Construct the diagonals of the rhombus and then use the software to measure the angles at the intersection of the two diagonals. What are the measures of the angles? Do the angle measures change as you drag a vertex?

   c. What does your answer to the previous question imply about the diagonals of a rhombus?

3. Construct a rectangle. Be sure that the figure remains a rectangle as you drag any of its vertices.

   a. Write a short summary of the steps you took to complete the construction.

   b. Construct the diagonals of the rectangle and then use the software to measure their lengths. What do you notice? Does your observation change as you drag a vertex?

   c. What does your answer to the previous question imply about the diagonals of any rectangle?

4. Construct a square. Be sure that the figure remains a square as you drag any of its vertices.

   a. Write a short summary of the steps you took to complete the construction.

   b. Because a square is a special kind of rhombus, what must be true about its sides and diagonals? Test your conjectures using the software.

   c. Because a square is a rectangle, what must be true about its angles and diagonals? Test your conjectures using the software.

Part C: Summary and Connections

1. What other properties of triangles and quadrilaterals can you explore using geometry software?

2. a. What are the advantages of exploring triangles and quadrilaterals with geometry software?

   b. How might you use geometry software to teach concepts about triangles and quadrilaterals in your classroom?

Exploration 11.12 Constructing Regular Polygons

Purpose: Students construct regular polygons using geometry software.

Materials: Geometry software

Part A: Constructing a Regular Hexagon

Use the segment tool to construct a segment and label its endpoints A and B. Next, place a point C anywhere on the screen and construct a circle with center C and radius $\overline{AB}$. Construct a point on the circle and label it D. Construct a circle with center D and radius $\overline{AB}$, and then construct the intersection of the two circles. Continue to construct circles using the new intersection points as the centers and $\overline{AB}$ as the radius until you have six points on the original circle. Construct segments to form the hexagon. Hide the circles, leaving only the hexagon.

1. a. Why does construction lead to a regular hexagon?
   b. What happens if you click and drag A or B? What about C? What about D?

2. Construct an equilateral triangle using three vertices from the regular hexagon. Write a summary of the steps you took to complete the construction. Color the triangle blue.

3. Construct a regular dodecagon (12-gon) using the regular hexagon and six other vertices. Write a summary of the steps you took to complete the construction. Color the dodecagon red. How did you know where to place the six other vertices to form the dodecagon?

Part B: Constructing a Regular Octagon

Clear the screen from the previous construction and then construct a line $\overleftrightarrow{AB}$. Place a point C anywhere on the screen and construct a line that is perpendicular to $\overleftrightarrow{AB}$ and passes through C. Label their intersection D. Bisect the four right angles created by the perpendicular lines. Construct a circle with center D and having any radius. Construct the eight points of intersection between the lines and the circle. Construct segments to form the octagon and then hide the lines and circles, leaving only the octagon.

1. a. Why does the construction lead to a regular octagon?
   b. What happens if you click and drag A or B? What about C? What about D?

2. What other regular polygons can you construct once you have a regular octagon?

Part C: Using Rotations to Construct Regular Polygons

The first two parts of the exploration used traditional constructions to construct regular polygons. Rotations can be used to construct other regular polygons. For instance, consider a regular pentagon. Clear the screen from the previous construction and then construct a segment $\overline{AB}$. Mark B as the center of rotation and then rotate $\overline{AB}$ and both endpoints by 108°. A rotated segment should appear. Mark the new endpoint as the center of rotation and rotate the segment again. Continue the process until you have completed the pentagon.

1. a. Why does the construction lead to a regular pentagon?
   b. What happens if you click and drag A or B?
   c. Why was 108° used as the angle of rotation?
   d. What other regular polygons can you construct once you have a regular pentagon?

2. Use rotations of a segment to construct a regular nonagon. How did you determine the measure of the angle of rotation?

3. a. Use rotations of a segment to construct a regular decagon. How did you determine the measure of the angle of rotation?
   b. How could you construct a regular decagon after your have constructed a regular pentagon?

4. Can any regular polygon be constructed by using rotations of a segment? If so, what information must be known to make the construction?

Part D: Summary and Connections

1. What properties of regular polygons can you explore using GSP?

2. Would Euclid have considered using rotations a legitimate way to make a construction? Why or why not?

Chapter 12
Coordinate and Transformation Geometry

Chapter 12 continues the study of shapes by considering coordinate and transformation geometry. In **coordinate geometry,** shapes are represented on the Cartesian coordinate system, making it possible to describe and manipulate them with algebraic expressions. In the classroom, children use coordinate geometry to improve their understanding of spatial relationships by answering questions about direction, distance, and location. Children also learn to use coordinate systems to analyze shapes, navigate to specific locations, and measure distances.

In **transformation,** or **motion geometry,** we describe the change in the position and orientation of a shape that results from simple motions like slides, flips, and turns. In the early grades, children learn about these motions by physically manipulating shapes. However, as their spatial reasoning skills improve, children learn to move objects mentally and to track the movements of an object on a coordinate system.

Transformations provide a key way to describe and understand symmetry. An object has **symmetry** if it can be transformed in some way back onto itself and still have its original appearance. Both transformations and symmetry are used to study geometric patterns like tessellations. **Tessellations** are patterns made from a set of basic figures that cover the entire plane with no overlaps or gaps. Not only do they provide an excellent opportunity for children to apply what they have learned about transformations and symmetry, but they also give students an opportunity to connect mathematics to many real-world situations.

The explorations in Chapter 12 investigate many facets of coordinate and transformation geometry. Specifically,

- Explorations 12.1 - 12.4 look at ways to represent shapes on a coordinate system and applications of coordinate geometry.
- Explorations 12.5 - 12.10 focus on understanding transformations and using them to describe symmetry.
- Explorations 12.11 - 12.13 consider ways to classify and make geometrical patterns.

Exploration 12.1 Taxicab Geometry

Purpose: Students use taxicab geometry to explore a different notion of length on the Cartesian coordinate plane.

Materials: Grid paper

Part A: Understanding Taxicab Length

Taxicab geometry is similar to coordinate geometry. The only difference between them is the way in which distance is measured. In coordinate geometry, the Euclidean distance between two points, $d_E(P, Q)$, is the length of a line segment connecting P and Q. In taxicab geometry, we think of the coordinate plane as a set of city streets that are arranged in a rectangular grid. We then imagine a taxicab that can only travel along the streets (the grid lines) and not diagonally across a block. The **taxicab distance** between two points, $d_T(P, Q)$, is the smallest number of blocks a taxicab must drive to get from P to Q. For instance, in Figure 1 the taxicab distance from A to B is $d_T(A, B) = 10$ units.

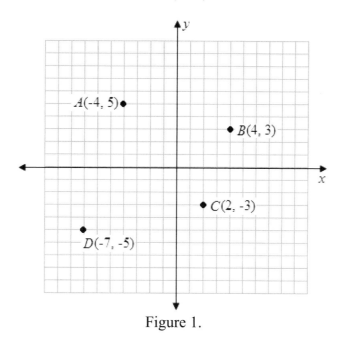

Figure 1.

1. a. Use Figure 1 to compute each taxicab distance.
    i. $d_T(A, C)$ ii. $d_T(B, C)$ iii. $d_T(C, D)$ iv. $d_T(B, D)$
   b. Use the distance formula to compute the Euclidean distance.
    i. $d_E(A, B)$ ii. $d_E(B, C)$ iii. $d_E(C, D)$ iv. $d_E(A, D)$
   c. Compare the taxicab distances to the Euclidean distances. What generalization can you make from your comparisons? Can you write your generalization symbolically?

2. Is the taxicab distance between two points unique? What about the path taken to determine the taxicab distance?

3. If $A = (x_1, y_1)$ and $B = (x_2, y_2)$, write a formula that gives the taxicab distance between points A and B.

4. Plot the point $C(4, 2)$ on grid paper. Find eight other points for which their taxicab distance from C is 6. Plot each point and give its coordinates.

5. a. If $d_T(A, B) = d_T(C, D)$, does it follow that $d_E(A, B) = d_E(C, D)$?
   b. If $d_E(A, B) = d_E(C, D)$, does it follow that $d_T(A, B) = d_T(C, D)$?
   c. What must be true about two points A and B if $d_T(A, B) = d_E(A, B)$?

Part B: Problems in Perfectia

Use grid paper to help you solve the following problems in Perfectia, a city in which the streets are laid out in a perfect rectangular grid.

1. Linda and Louie are looking for an apartment in Perfectia. Linda works for a school at $A(-4, -2)$ and Louie plans to open a small business at $B(2, 2)$. They decide that the distance Linda travels to work plus the distance that Louie travels to work should be as small as possible. Where should they look to live?

2. A fire breaks out at a home located at $H(3, -3)$. The two closest fire stations are located at $E(5, 2)$ and $D(1, -1)$. Which station should the dispatcher contact first to send fire engines to the home?

3. Devon's car runs out of gas at $D(4, 4)$. There are three gas stations located at $G(4, 7)$, $H(3, 2)$, and $I(6, 3)$. Which gas station is closest?

4. There are three elementary schools in Perfectia with locations at $A(-3, 5)$, $B(1, 1)$ and $C(-1, -4)$. Draw boundary lines between the schools so that each student in Perfectia attends the closest school.

Part C: Summary and Connections

1. Which do you think is a more realistic measure of distance: Euclidean distance or taxicab distance? Explain your thinking.

2. How might you use taxicab geometry to teach spatial reasoning to a group of middle grade students?

Exploration 12.2 Coordinate Geometry on a Geoboard

Purpose: Students use a geoboard to investigate coordinate geometry.

Materials: 11 × 11 geoboard and rubber bands

Part A: Points on a Geoboard

1. Consider the following 11 × 11 geoboard in which the peg in the lower left corner has coordinates (0, 0).
   a. Give the coordinates of the labeled points.
   b. What quadrant of the Cartesian plane does the geoboard represent?

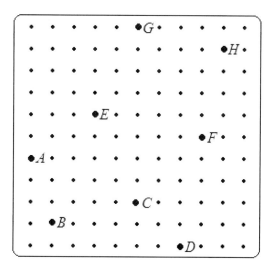

2. Repeat Question 1, but this time assume the peg in the lower right corner has coordinates (0, 0).

3. Repeat Question 1, but this time assume the peg in the upper left corner has coordinates (0, 0).

4. Assume the peg in the lower left corner of your geoboard has coordinates (0, 0).
   a. Locate each set of points and join them in the order given. What word is formed by the three sets of points?
      i. (0, 5), (0, 9), (1, 9), (2, 8), (2, 6), (1, 5), (0, 5)
      ii. (4, 3), (4, 7), (6, 7), (6, 3), (4, 3)
      iii. (9, 5), (7, 5), (7, 1), (9, 1), (9, 3), (8, 3)
   b. Create another small word on the board. Write a set of coordinates, similar to those in part (a), which can be used to generate your word. Give the coordinates to a partner to re-create your word.

Part B: Segments on a Geoboard

Let the peg in the lower left corner have coordinates (0, 0). Construct the segment $\overline{AB}$ with endpoints $A(1, 2)$ and $B(4, 1)$.

1. Consider the point $C(2, 5)$. Find a point D so that $\overline{CD}$ is parallel to $\overline{AB}$.

2. Find another pair of points that creates a segment parallel to $\overline{AB}$ but is twice as long. Give the coordinates of the endpoints and explain how you know the segments are parallel.

3. Consider the point $E(3, 2)$. Find a point F so that $\overline{EF}$ is perpendicular to $\overline{AB}$.

4. Find another pair of points that creates a segment perpendicular to $\overline{AB}$ but is twice as long. Give the coordinates of the endpoints and explain how you know the segments are perpendicular.

Part C: Geoboard Battleship

Get together with one of your peers and get two 11 × 11 geoboards to play geoboard battleship. Begin the game by deciding what kind of polygon will serve as the ship in the first round. Player 1 creates a ship on a geoboard, and player 2 tries to sink the ship by guessing its exact location.

Player 2 calls out "shots" by specifying the whole-number coordinates of points on the geoboard, where the peg in the lower left corner has coordinates (0, 0). Player 1 must specify whether the shot is a "hit" or a "miss." If the shot is a hit, Player 1 should specify whether the shot is *within* the ship, at a *vertex*, or on the *perimeter* of the ship. Player 2 should keep an organized record of each shot and how many shots are made.

The goal is to discover the location of the ship in the fewest number of shots. When ready to identify the location of Player 1's ship, Player 2 ends the round by constructing the ship and comparing it to Player 1's geoboard. Add a shot to Player 2's total for each incorrect vertex. After counting shots, the players switch roles and play again. After each player has taken a turn, the winner is the player with the fewest number of shots.

Part D: Summary and Connections

1. What are the advantages and disadvantages of using a geoboard rather than graph paper to complete this exploration?

2. How can you adapt this exploration for use in an elementary classroom? Specifically, what parts would you use and how would you simplify or extend them?

Exploration 12.3 Coordinate Geometry and Polygons

Purpose: Students use a geoboard to create and explore polygons.

Materials: 11 × 11 geoboard and rubber bands

Part A: Triangles on a Geoboard

Use an 11 × 11 geoboard to complete each activity. Let the peg in the lower left corner have coordinates (0, 0).

1. For each set of three vertices, determine whether the triangle they represent is acute, right, or obtuse and then scalene, isosceles, or equilateral.
    a. $A(1, 1)$, $B(4, 4)$, and $C(6, 2)$ b. $A(6, 4)$, $B(4, 8)$, and $C(8, 10)$

2. Construct a segment $\overline{AB}$ with endpoints $A(0, 0)$ and $B(0, 8)$. Find coordinates for a point C to make $\triangle ABC$ into each of the following types of triangle.
    a. Right triangle.   b. Isosceles triangle.   c. Right, isosceles triangle.

3. Construct a segment $\overline{WX}$ with endpoints $W(3, 4)$ and $X(6, 5)$.
    a. Find the coordinates of point Y so that $\triangle WXY$ is a right triangle with leg $\overline{WX}$.
    b. Find at least one other solution to part a.
    c. Find the coordinates of point Y so that $\triangle WXY$ is a right triangle with hypotenuse $\overline{WX}$.

4. Is it possible to create an equilateral triangle on a geoboard? If so, give possible coordinates for the three vertices. If not, explain why not.

Part B: Quadrilaterals on a Geoboard

Use an 11 × 11 geoboard to complete each activity. Let the peg in the lower left corner have coordinates (0, 0).

1. Construct a segment $\overline{AB}$ with endpoints $A(6, 4)$ and $B(5, 6)$. Find the coordinates of all points C and D so that quadrilateral $ABCD$ is a
    a. square.   b. rectangle with two sides twice the length of $\overline{AB}$.

2. Locate the points $A(5, 6)$, $B(6, 4)$, and $C(8, 5)$. Find all possible coordinates for a point D, so that A, B, C, and D form a rhombus.

3. Construct a parallelogram that has a 45° angle and is not a rhombus.
   a. What are the coordinates of its vertices?
   b. Can you make two other parallelograms that satisfy the same criteria?
   c. Are the three parallelograms you have constructed similar? Explain your thinking.

4. Construct a trapezoid $ABCD$ so that $\overline{AB} \parallel \overline{CD}$ and sides $\overline{AB}$ and $\overline{CD}$ are an even number of units long. Construct the diagonals of the trapezoid. Next, locate the midpoints of $\overline{AB}$ and $\overline{CD}$, and construct the segment between them. Call the new segment $\overline{EF}$.
   a. What is true about $\overline{EF}$ in relationship to the diagonals of the trapezoid?
   b. Test your conjecture with other trapezoids that satisfy the given conditions. Does your conjecture always hold true? Can you prove it?

Part C: Regular Polygons on a Geoboard

Use an 11 × 11 geoboard to complete each activity. Let the peg in the lower left corner have coordinates (0, 0).

1. Is it possible to create a regular pentagon on a geoboard? If so, give possible coordinates for the five vertices. If not, explain why not.

2. Is it possible to create a regular hexagon on a geoboard? If so, give possible coordinates for the six vertices. If not, explain why not.

3. a. What regular polygons can be constructed on a geoboard?
   b. What are the limiting factors of a geoboard that prevent many regular polygons from being constructed?

Part D: Summary and Connections

1. a. How could you adapt the way the geoboard is used in this activity so that negative coordinates are used?
   b. How could you use four geoboards to better represent the Cartesian coordinate plane?

2. Design an activity for elementary students that uses a geoboard to teach a property of triangles.
   a. What features of the geoboard did you use that would make your activity a meaningful learning experience for elementary students?
   b. What disadvantages do your foresee in using a geoboard in an elementary classroom? What could you do to help your students overcome these disadvantages?

Exploration 12.4 Making a Coordinate Map

Purpose: Students use grid paper to make coordinate maps of their surroundings.

Materials: Half-inch grid paper, a tape measure or other measuring tools, and tape

Part A: Mapping Your Classroom

Form a small group with two or three of your peers. Use half-inch grid paper to make a coordinate map of your classroom. If necessary, tape several sheets of grid paper together to make a sheet large enough to make your map. Your map should include the location of all items in the room, such as the desks, the door, the trash can, the black board, the windows, and so on. When necessary, use a tape measure to measure unknown distances. Pick a point on your map to serve as the origin, that is, to have the coordinates (0, 0). After your map is complete, use it to answer each question.

1. What point did you choose to serve as the origin? Why did you choose this point?

2. What is the scale of your map? Why did you choose this scale?

3. Consider the location of your desk relative to the origin.
    a. What are the coordinates of your desk relative to the origin?
    b. Which makes more sense: to describe your desk as a single point or as a set of points? Explain your thinking.

4. How would you describe the relative position of the chalkboard to the origin? Are there any difficulties in describing the position of the board with coordinates? How did you work around these difficulties?

5. Write a set of directions using coordinates that describe how to get from your desk to the door so that anyone following the directions would not bump into any obstacles.

Part B: Mapping Your Campus

Use half-inch grid paper to make a coordinate map of some or all of your college campus. If necessary, tape several sheets of grid paper together to make a sheet that is large enough to make your map. Your map should include the buildings on your campus, as well as other important features like parking lots, recreation fields, and so on. Even though you may not know exact distances, do your best to estimate the positions of the major features relative to one another. Place a compass on your map to indicate the directions of north, south, east and west. When your map is complete, use it to answer each question.

1. What is the scale of your map? Why did you choose this scale?

2. Use approximate measures and directions like north, south, east, and west to describe the relative position of your classroom building to the three buildings that are closest to it.

3. Write a set of directions using approximate measures and directions like north, south, east, and west to describe how you would walk to get from your classroom building to your car or dorm.

4. Write a set of directions using approximate measures and directions like north, south, east, and west that describe how a person would walk to get from your classroom building to the building on your map that is farthest from it.

Part C: Summary and Connections

1. Which do you think is more difficult: to give directions using coordinates or to give directions using north, south, east, and west?

2. Why do you think most map makers put a coordinate system on their maps?

3. How might you adapt this exploration for use in an elementary classroom? Specifically, what parts would you use and how would you simplify or extend them?

Exploration 12.5 Making Transformations

Purpose: Students use concrete materials to explore translations, rotations, and reflections.

Materials: Patty paper, plastic reflector, protractor, ruler, compass, and quarter-inch grid paper

Part A: Making Translations

To fully understand geometry, we must learn how shapes can be moved and how to represent those motions. We call such motions **transformations,** or **rigid motions**. In the elementary classroom, three rigid motions are often explored: translations, rotations, and reflections.

A **translation**, or **slide**, is a rigid motion that takes an object from one location to another without twisting or turning it. To make a translation, the direction and the length of the translation must be known.

1. Draw and label a figure like the one shown on a sheet of quarter-inch grid paper. Translate the figure in each way. Once you have completed each translation, describe the process you used to do so.

   a. To the right 1 in. b. Down $\frac{3}{4}$ in. c. To the left $1\frac{1}{2}$ in.

   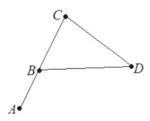

2. The direction and the length of a translation are typically given by a translation vector. In each case, describe the motion indicated by the translation vector and then make the translation. Describe the process you used to do so.

   a.    b.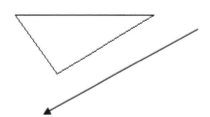

Part B: Making Rotations

Another rigid motion is a **rotation** or **turn**. To make a rotation, the center of rotation and the turn angle must be known. The center of rotation is the point about which the shape is rotated. It can be any point that is in, on, or outside the shape. The turn angle can be clockwise or counterclockwise and an angle measure is given to indicate how much of a rotation is to be made.

1. Draw a triangle on a sheet of patty paper, and label its vertices *A*, *B*, and *C*. Place three points on the patty paper; a point *P* inside the triangle, a point *Q* on the triangle, and a point *R* outside the triangle. First, predict where you believe the image of the shape will be under the given rotation. Then, use your protractor, ruler, and compass to make the rotation.
    a. Clockwise by 90° about *R*
    b. Clockwise by 60° about *P*
    c. Counterclockwise by 120° about *Q*
    d. Counterclockwise by 45° about *C*

Part C: Making Reflections

A third rigid motion is a **reflection,** or a **flip**. To make a reflection, a line called the axis of reflection must be known. The axis of reflection acts as a mirror that takes every point on the shape and maps it to a unique point on the opposite side of the axis of reflection and at the same distance. Any point on the axis of reflection remains fixed.

1. Draw a nonsymmetrical quadrilateral on a sheet of paper and label its vertices *A*, *B*, *C*, and *D*. Place three lines on the paper: a horizontal line *l*, a vertical line *m*, and a line *n* that is neither horizontal nor vertical. First predict where you believe the image of the shape will be under the reflection. Then, use your plastic reflector and ruler to make the reflection.
    a. A reflection in line *l*
    b. A reflection in line *m*
    c. A reflection in line *n*
    d. A reflection in side $\overline{AD}$

Part D: Summary and Connections

1. Is it easiest for you to visualize the image of a shape under a reflection, a translation, or a rotation? Why?

2. How might grid paper or a coordinate system be useful in making translations, rotations, and reflections?

Exploration 12.6 Exploring Transformations with Geometry Software

Purpose: Students use geometry software to explore translations, rotations, and reflections.

Materials: Geometry software

Part A: Exploring Translations

Use geometry software to create a non-symmetrical polygon and its interior like the one shown.

1. Construct a segment $\overline{AB}$. Use the software to translate the figure by $\overline{AB}$. Change the color of the interior of the translated image.
    a. Describe the procedure you used to translate the figure by $\overline{AB}$.
    b. What happens to the image as you drag B?
    c. Why is it necessary to provide the software with both a distance and a direction for the translation? During what part of your procedure did you indicate the distance and direction of the translation?
    d. Construct a segment that connects a vertex on the original shape to its image. Make the new segment dashed. How does the new segment compare to the translation vector $\overline{AB}$?

2. a. Are there other ways to use the software to make a translation? If so, describe how the other procedures are different from the one you used in Question 1.
    b. What are the advantages and disadvantages of each method for translating the shape? Which do you think is the better method?

Part B: Exploring Rotations

Clear your screen and create another non-symmetrical polygon and its interior.

1. Construct an angle $\angle ABC$. Use the software to rotate the polygon by $\angle ABC$. Change the color of the interior of the rotated image.
    a. Describe the procedure you used to rotate the polygon by $\angle ABC$.
    b. What happens to the image as you drag A? Drag B? Drag C?

c. Why is it necessary to provide the software with both a center of rotation and a turn angle to make the rotation? During what part of your procedure did you indicate the amount and direction of the rotation?

2. a. Are there other ways to use the software to make a rotation? If so, describe how the other procedures are different from the one you used in Question 1.
   b. What are the advantages and disadvantages of each method for rotating the shape? Which do you think is the better method?

Part C: Exploring Reflections

Clear your screen and create another non-symmetrical polygon and its interior. Construct a line $\overleftrightarrow{XY}$ anywhere on the screen. Use the software to reflect the polygon in line $\overleftrightarrow{XY}$. Change the color of the interior of the reflected image.

1. Describe the procedure you used to reflect the polygon in $\overleftrightarrow{XY}$.

2. Drag the polygon. What is the relationship between the polygon and its image with respect to $\overleftrightarrow{XY}$?

3. Drag X or Y. What happens?

4. Why is it necessary to provide the software with an axis of reflection to make the reflection? During what part of your procedure did you designate this line?

Part D: Summary and Connections

1. A **glide reflection** is a translation followed by a reflection. Create a shape and its interior. Use geometry software to transform the shape by a glide reflection.
   a. Summarize the steps you took to make the glide reflection.
   b. What information did you have to provide to the software to make the glide reflection? During what part of your procedure did you designate this information?

2. Create a shape and its interior, and then construct two parallel lines. Label them m and n. Use geometry software to make two consecutive reflections of your shape, first in line m and then in line n.
   a. Summarize the steps you took to complete the transformation.
   b. How does the final reflected image compare to the original shape? Is there a single transformation that is equivalent to the reflection in two parallel lines?
   c. Do you get the same result if you first reflect in line n and then in line m?

3. What greater understandings did you gain by exploring geometric transformations with geometry software?

Exploration 12.7 Spatial Reasoning and Transformations

Purpose: Students use spatial reasoning to identify transformations.

Materials: Paper and pencil

Part A: Translations

1. In each case, identify the translation that maps △ABC onto △A'B'C'.

   a.
   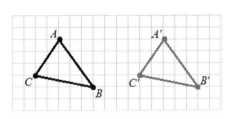

   b.
   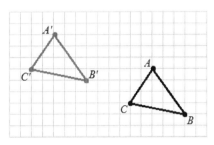

   c.
   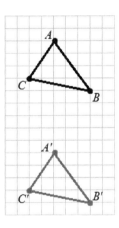

   d.
   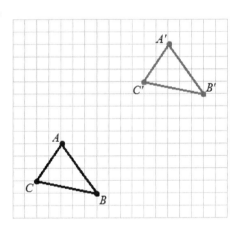

2. What information did you need to give to identify a particular translation?

Part B: Rotations

1. In each case, identify the rotation that maps △ABC onto △A'B'C'.

   a.
   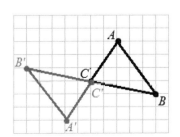

   b.
   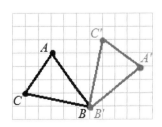

c.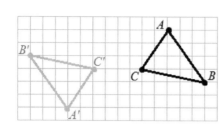

d.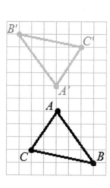

2. What information did you need to give to identify a particular rotation?

Part C: Reflections

1. In each case, identify the reflection that maps $\triangle ABC$ onto $\triangle A'B'C'$.

   a.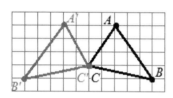

   b.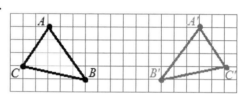

   c.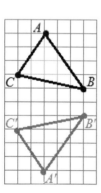

   d.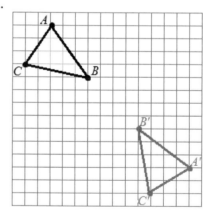

2. What information did you need to give to identify a particular reflection?

Part D: Summary and Connections

1. a. Why is a grid system useful when identifying transformations?
   b. How could you identify a translation if a grid system is not given? A rotation? A reflection?

2. a. How did you use spatial reasoning to complete this activity?
   b. What other aspects of spatial reasoning can be taught and learned through transformations?

Exploration 12.8 Size Transformations

Purpose: Students use measurement tools to explore size transformations.

Materials: Ruler, protractor, colored pencils, and paper

Part A: Dilating a Quadrilateral

In addition to rigid motions, such as translations, rotations, and reflections, there are also size transformations. A **size transformation**, or **dilation**, preserves the shape of a figure, but not necessarily its size. To make a size transformation, two things must be known: the **center**, which is the point of origin of the size transformation, and the **scale factor**, which tells how much the figure is increased or decreased.

For instance, consider a dilation of quadrilateral *ABCD* by a scale factor of 2. Begin by making a copy of the quadrilateral and placing a point *P* relatively close to, but outside of, the quadrilateral.

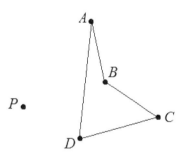

Use your ruler to locate and label the point *A'* so that $\overline{PA'}$ is twice the length of $\overline{PA}$. In the same way locate and label points *B'*, *C'*, and *D'*. Use a red colored pencil to draw segments $\overline{A'B'}$, $\overline{B'C'}$, $\overline{C'D'}$, and $\overline{D'A'}$ to create quadrilateral *A'B'C'D'*.

1. Measure the sides of quadrilateral *ABCD* and quadrilateral *A'B'C'D'*. Compare the measures. What do you notice about the lengths of the corresponding sides?

2. a. Consider the location of *A'* relative to *P* and *A*. What is true about the three points? Is this true for other vertices and their images?
   b. Based on your observations in the previous question, what is true about the center of a dilation, a point, and the image of the point?

Part B: The Role of the Center

Using the same quadrilateral *ABCD*, select a different point *K* to serve as the center. Make *K* outside the quadrilateral and slightly farther away from the quadrilateral than *P*. Find the image of quadrilateral *ABCD* under the size transformation with *K* as the center and a scale factor of 2. Use a blue colored pencil to draw the image.

1. Compare the red image to the blue image. How did changing the center of dilation affect
    a. the direction of the image? b. the distance of the image from quadrilateral *ABCD*?

2. How did the distance of the center of dilation from quadrilateral *ABCD* affect the size of the image?

3. What do you think would happen if the center of dilation was inside the quadrilateral? On the quadrilateral? Test your conjecture to see if you are correct.

4. Summarize how a change in the center affects the dilation.

Part C: The Role of the Scale Factor

Find the image of quadrilateral *ABCD* under the size transformation with *P* as the center of dilation and a scale factor of 3. Use a green colored pencil to draw the image.

1. Compare the red image to the green image. How did changing the scale factor affect
    a. the location of the image?   b. the size of the image?

2. Draw another image of quadrilateral *ABCD* with *P* as the center and a scale factor of $\frac{1}{2}$.

    Use an orange colored pencil to draw the image. How did changing the scale factor affect
    a. the location of the image?   b. the size of the image?

3. What would be the result of a size transformation if the scale factor was 1?

4. Summarize how a change in the scale factor affects the dilation.

Part D: Summary and Connections

1. Is a dilation an isometry? Why or why not?

2. Use geometry software to explore dilations electronically.
    a. What is the advantage of using software to make size transformations?
    b. Write a set of directions explaining how to use software to make a size transformation with a scale factor of 2.

Exploration 12.9 Symmetry

Purpose: Students explore the symmetries of geometrical shapes.

Materials: Pattern blocks

Part A: Reflection Symmetry

A planar shape has **reflection symmetry** if the figure can be "folded" so that one side lies exactly on top of the other. The fold line forms the **axis** or **line of symmetry**, which is equivalent to an axis of reflection. Consequently, a figure has reflection symmetry if and only if there is a line *l* such that a reflection in *l* maps the figure onto itself.

1. Draw all lines of symmetry for each shape.

   a.   b.   c.   d.

2. A standard set of pattern blocks has six shapes: a regular hexagon, an isosceles trapezoid, two rhombi, a square, and an equilateral triangle. Describe the reflection symmetry for each shape.

3. Draw a planar shape with each of the given symmetries.
   a. A vertical line of symmetry
   b. A horizontal line of symmetry
   c. A vertical line and a horizontal line of symmetry
   d. No reflection symmetry

4. Draw a planar shape with the given number of lines of symmetry.
   a. 1   b. 2   c. 3   d. 4   e. 5   f. 10

5. Can a planar shape have more than one vertical line of symmetry? If so, draw such a shape. If not, explain why not.

Part B: Rotation Symmetry

A planar shape has **rotation symmetry** if the shape can be mapped onto itself by rotating it around a point through a turn angle less than 360°. If a figure has a rotation symmetry of α degrees, then it also has a rotation symmetry of $n\alpha$ degrees, where *n* is a natural number. The rotation symmetry is named using the turn angle with the smallest measure. For instance, a square is said to have 90°-rotation symmetry, even though it also has 180°- and 270°-rotation symmetry.

1. Describe the rotation symmetry for each polygon.
   a.   b.   c.   d.

2. Describe the rotation symmetry for each shape in a standard set of pattern blocks.

3. Draw a planar shape with rotation symmetry for the given turn angle.
   a. 180° b. 120° c. 90° d. 60° e. 45° f. 30°

4. Can a planar shape have rotation symmetry without reflection symmetry? If so, draw such a figure. If not, explain why not.

5. Does a circle have rotation symmetry? Explain.

Part C: Summary and Connections

1. Take a few moments to walk around your classroom building or around your campus. As you walk, look for shapes with symmetry. Specifically, find three shapes with the specified symmetries. Draw a diagram of each shape and indicate the symmetry.
   a. Reflection symmetry but no rotation symmetry
   b. Rotation symmetry but no reflection symmetry
   c. Reflection symmetry and rotation symmetry
   d. No symmetry

2. Why is symmetry an important aspect of geometrical shapes?

Exploration 12.10 Defining Shapes with Symmetry

Purpose: Students use symmetry to write alternative definitions for planar shapes.

Materials: Paper and pencil

Part A: Redefining Triangles

Different planar shapes generally have different reflection and rotation symmetries. Because this is true, symmetry can be used to classify and define them. For instance, consider triangles.

1. Draw a scalene, an isosceles, and an equilateral triangle. Investigate their reflection and rotation symmetries and record what you find.
    a. How do the symmetries differ for each type of triangle?
    b. Use symmetry to write an alternative definition for each type of triangle. For instance, a scalene triangle can be defined as a triangle with no reflection and no rotation symmetries.

2. Draw an acute, a right, and an obtuse triangle. Investigate their reflection and rotation symmetries and record what you find.
    a. How do the symmetries differ for each type of triangle?
    b. If possible, use symmetry to write an alternative definition for each type of triangle. If it is not possible, explain why.

Part B: Redefining Quadrilaterals

Investigate the reflection and rotation symmetries for each type of quadrilateral. If possible, use symmetry to write an alternative definition for each one. If it is not possible, explain why.

    1. Trapezoid   2. Isosceles Trapezoid   3. Parallelogram
    4. Rhombus   5. Rectangle   6. Square

Part C: Summary and Connections

1. What are the advantages and disadvantages of defining triangles and quadrilaterals with symmetry?

2. Investigate the reflection and rotation symmetries of regular polygons.
    a. If possible, use symmetry to write an alternative definition for a regular pentagon, hexagon, and heptagon. How are the definitions similar?
    b. If possible, use symmetry to write an alternative definition for a regular n-gon.

Exploration 12.11 Border Patterns

Purpose: Students use pattern blocks to make and classify border patterns.

Materials: Pattern Blocks

Part A: Making Border Patterns

A **border pattern** is created when a basic design, called a **motif**, is repeated in a linear fashion through multiple iterations of a translation or a glide reflection.

1. Use of a set of pattern blocks to create two or three border patterns from iterations of a translation. Make at least one pattern have more than one shape in the motif. In each case, record the pattern and identify the motif.
    a. Why must the motif of the border patterns occur at regular intervals?
    b. What determines the direction of the border pattern?
    c. What must be true about the orientation of the motif throughout the pattern? Why is this true?

2. Use a set of pattern blocks to create two or three border patterns from iterations of a glide reflection. Make at least one pattern have more than one shape in the motif. In each case, record the pattern and identify the motif.
    a. Why must the motif of the border patterns occur at regular intervals?
    b. What determines the direction of the border pattern?
    c. How does the glide reflection affect the appearance of the border pattern?

Part B: Border Patterns and Symmetry

Border patterns can have symmetry because they are generated from a single motif. However, because they are infinite in nature, any symmetry in the pattern must map the entire infinite pattern back onto itself.

1. Some border patterns have **reflection symmetry**, which means there is a line *l* such that a reflection of the border pattern in *l* maps it onto itself.
    a. If possible, use pattern blocks to create each border pattern. Record the pattern, identify the motif, and find at least one line of symmetry.
        i. A border pattern with vertical, but not horizontal reflection symmetry.
        ii. A border pattern with horizontal, but not vertical reflection symmetry.
        iii. A border pattern with both vertical and horizontal reflection symmetry.
        iv. A border pattern with no reflection symmetry.
    b. Compare the reflection symmetry in each border pattern with the reflection symmetry in its motif. What do you notice?

2. Border patterns can also have **rotation symmetry**, which means the border pattern can be mapped onto itself by rotating it around a point through a turn angle less than 360°. Use pattern blocks to create two or three border patterns with rotation symmetry. In each case, record the pattern, identify the motif, and find at least one turn center.
    a. Is the turn center of a border pattern always a turn center of the motif?
    b. Compare the rotation symmetry of each border pattern with the rotation symmetry of its motif. Are they the same? If not, describe how they differ?

3. Some border patterns have **translation symmetry**, which means there is a translation that maps the pattern back onto itself. Use pattern blocks to create two or three border patterns with translation symmetry. In each case, record the pattern, identify the motif, and identify one translation vector. Will every border pattern have translation symmetry? Why or why not?

4. Border patterns can have **glide reflection symmetry**, which means there is a glide reflection that maps the pattern back onto itself. Use pattern blocks to create two or three border patterns with glide reflection symmetry. In each case, record the pattern, identify the motif, and identify one translation vector and the axis of reflection.
    a. Will every border pattern have glide reflection symmetry? Why or why not?
    b. Will every border pattern with glide reflection symmetry also have translation symmetry? Why or why not?

5. If possible, make a border pattern with the given symmetries.
    a. Vertical reflection symmetry and glide reflection symmetry
    b. Horizontal reflection symmetry and glide reflection symmetry
    c. Rotation symmetry and vertical reflection symmetry
    d. Rotation symmetry and glide reflection symmetry
    e. Glide reflection symmetry with no horizontal reflection symmetry
    f. Rotation symmetry without horizontal reflection symmetry
    g. No reflection symmetry but has glide reflection symmetry
    h. No symmetry but has translation symmetry

6. If you were to classify border patterns by their symmetry, how many types of border patterns would there be? Can you make one of each type?

Part C: Summary and Connections

1. Find five border patterns in your classroom or classroom building. Record the patterns and describe the symmetry in each.

2. What other manipulatives can you use to make border patterns?

Exploration 12.12 Wallpaper Patterns

Purpose: Students use pattern blocks to make and explore wallpaper patterns.

Materials: Pattern Blocks

Part A: Making Wallpaper Patterns

A **wallpaper pattern** is created when a basic design, called a **motif**, is repeated in two directions through multiple iterations of a translation or a glide reflection.

1. Use a set of pattern blocks to create two or three wallpaper patterns through iterations of a vertical and a horizontal translation. Make at least one pattern have more than one shape in the motif. Record each pattern and identify the motif.
    a. Why does the motif occur at regular intervals?
    b. What is true about the orientation of the motif throughout the pattern?

2. Use a set of pattern blocks to create two or three wallpaper patterns through iterations of a glide reflection and a vertical translation. Make at least one pattern have more than one shape in the motif. Record each pattern and identify the motif.
    a. Why does the motif occur at regular intervals?
    b. How does the glide reflection affect the appearance of the wallpaper pattern?

Part B: Wallpaper Patterns and Symmetry

Wallpaper patterns often have symmetry. However, because they are infinite, any symmetry must map the entire pattern back onto itself.

1. Some wallpaper patterns have **reflection symmetry**, which means there is a line l such that a reflection of the wallpaper pattern in l maps it onto itself. If possible, use pattern blocks to create each wallpaper pattern. Record the pattern, identify the motif, and find at least one line of symmetry.
    a. A wallpaper pattern with vertical, but no horizontal reflection symmetry.
    b. A wallpaper pattern with horizontal, but no vertical reflection symmetry.
    c. A wallpaper pattern with both vertical and horizontal reflection symmetry.

2. Wallpaper patterns can also have **rotation symmetry**, which means the pattern can be mapped onto itself by rotating it around a point through some turn angle less than 360°. Use pattern blocks to create two or three wallpaper patterns with rotation symmetry. Record the pattern, identify the motif, and find at least one turn center.
    a. Compare the rotation symmetry of each wallpaper pattern with the rotation symmetry of its motif. Are they always the same? If not, how do they differ?
    b. Is the turn center of the pattern always a turn center of the motif?
    c. What different rotation symmetries are possible in a wallpaper pattern?

3. Some wallpaper patterns have **translation** or **glide reflection symmetry**, which means there is a translation or glide reflection that maps the pattern back onto itself. Use pattern blocks to create two or three wallpaper patterns with such symmetry. Record the pattern, identify the motif, and identify the translation or glide reflection. Does every wallpaper pattern have translation or glide reflection symmetry? Why or why not?

4. Two wallpaper patterns are different types, only if their symmetry is different. How many different types of wallpaper patterns can you make using a set of pattern blocks? Be sure to consider reflection, rotation, translation, and glide reflection symmetries.

Part C: Summary and Connections

1. Find five wallpaper patterns in your classroom or classroom building. Record the patterns and describe the symmetry in each.

2. How are wallpaper patterns similar to tessellations? How are they different?

Exploration 12.13 Making Tessellations

Purpose: Students use manipulatives to create tessellations.

Materials: Pattern blocks, cardstock, grid paper, plain paper, ruler, scissors, and colored pencils

Part A: Making Simple Tessellations

A **tessellation**, or a **tiling** is a geometric pattern in which the motif fills the plane without overlaps or gaps. Motifs can be made from one shape or multiple shapes.

1. A standard set of pattern blocks has a regular hexagon, a trapezoid, two rhombi, a square, and an equilateral triangle.
    a. If only one shape is used, which pattern blocks will tessellate the plane?
    b. If possible, find a motif that uses two shapes and tessellates the plane.
    c. If possible, find a motif that uses three shapes and tessellates the plane.

2. A pentomino is a two-dimensional shape composed of five squares. Six of the twelve pentominoes are shown. Using only one shape at a time, colored pencils, and grid paper, determine whether each of the given pentominoes tessellates the plane. For those that do, what transformations of the pentomino did you have to use?

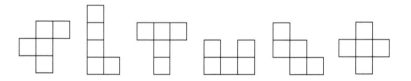

3. Use a ruler to make a small triangle on a sheet of cardstock and cut it out. Try to use your triangle to tessellate the plane by tracing copies of it on a sheet of unlined paper.
    a. If your triangle tessellated the plane, what transformations did you use to make the tessellation?
    b. How many copies of the triangle did you make before you decided whether or not it tessellated the plane?
    c. Do you think any triangle will tessellate the plane? Why or why not.

4. Use a ruler to make a small quadrilateral on a sheet of cardstock and cut it out. Try to use your quadrilateral to tessellate the plane by tracing it on a sheet of unlined paper.
    a. If your quadrilateral tessellated the plane, what transformations did you use to make the tessellation?
    b. How many copies of the quadrilateral did you make before you decided whether or not it tessellated the plane?
    c. Do you think any quadrilateral will tessellate the plane? Explain your thinking.

Part B: Making Escher-Like Drawings

Tessellations become more interesting when transformations are used to change the basic motifs used to make the tessellation.

1. One method, called the **cut-and-slide method**, uses translations to change the figure. Draw a 2 × 2-in. square on cardstock and cut it out. Draw designs on two adjacent sides of the square so that each design only touches the side at the beginning and the end and it does not go more than half way through the square. Carefully cut out the piece and translate the piece to the same position on the opposite side of the square. Tape the pieces together. Repeat the process for the design on the side.

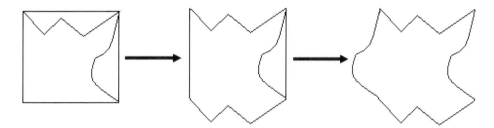

   Use your shape to tessellate the plane by tracing copies of it on a sheet of plain paper. Color your design.
   a. What transformations did you use to tessellate the plane with your shape?
   b. Can you explain why the cut-and-slide method works?
   c. What other basic shapes can you use with the cut-and-slide method to make a tessellation?

2. Another method is the **cut-and-turn** method. Draw a 2 × 2-in. square on cardstock and cut it out. Begin at a point on the top of the square and draw a jagged, random design to a point on the bottom of the square. Draw a second jagged line from the left to the right so that it intersects the first design. Starting from the left and proceeding clockwise, label the vertices of the square P, A, R, and T. Cut out the four pieces and arrange them so that the vertices meet at a point in the clockwise order of T, R, A, and P. Tape the pieces together.

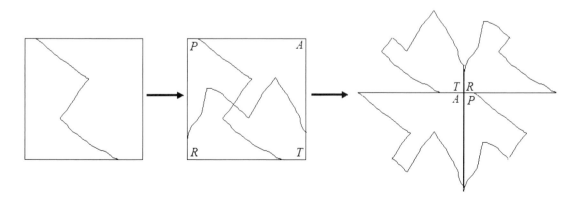

   Use your shape to tessellate the plane by tracing copies of it on a sheet of plain paper. Color your design.

a. What transformations did you use to tessellate the plane with your shape?
b. Can you explain why the cut-and-turn method works?
c. What other basic shapes can you use with the cut-and-turn method to make a tessellation?

Part C: Summary and Conclusions

1. What mathematical ideas are used to create tessellations?

2. Create a tessellation using the directions in Part A, but instead of beginning with a square begin with a regular hexagon.

3. Search the Internet to find websites that allow you to create tessellations electronically. Use the website to create several tessellations. What methods were used to create the tessellations?

4. M.C. Escher was an artist who was known for his use of tessellations. Search the Internet to see some of Escher's artwork.

Chapter 13
Measurement

Measurement is a topic often associated with geometry, not only because it gives a concrete way to characterize and study shapes, but also because geometric shapes have measurable attributes that are easy to recognize. Despite the close connection, measurement has several facets that are independent of geometry, making it a valuable topic of study in its own right.

Measurement is a process that can be learned, and children in the elementary grades spend several years coming to master it. They begin by learning to identify measurable attributes such as length, perimeter, area, volume, weight, temperature, and time. As they learn about these attributes, children are also introduced to the units used to measure them. Initially, children work with nonstandard units, a practice that not only helps them understand the measurement process, but also allows them to understand the need for standard units. They then move to measuring with both English and metric units. To make measures, students learn several measurement strategies such as making estimates, using measurement tools, and using formulas.

The explorations in Chapter 13 take an in-depth look at the mathematics involved in each step of the measurement process. Specifically,

- Explorations 13.1 - 13.4 consider measures of length and the units used to measure length.
- Explorations 13.5 - 13.14 focus on measurable attributes of geometric shapes such as perimeter, area, surface area, and volume.
- Explorations 13.15 - 13.16 explore other measurable attributes such as weight, time, and temperature.

Exploration 13.1 Nonstandard Units and Measures of Length

Purpose: Students measure lengths with nonstandard units and then examine the relationship between the size of the unit and the size of the measure.

Materials: Items having a length such as paper clips, coffee stirrers, pencils, etc.

Part A: Measuring Lengths with Nonstandard Units

When measuring lengths, any object with a length can be used as both the unit and the measurement tool. Such units are called non-standard because they vary from situation to situation. Measure the length of three objects in your classroom using two different units of measure. Record the measured objects and their lengths to the nearest whole unit and then use them to answer each question.

Object	Length with unit 1	Length with unit 2

1. Describe how you used the units to obtain your measures.

2. Which unit gave you the most accurate measure? Why do you think this is the case?

3. Look at the measures you obtained for each object. How did the size of the unit affect the measure of the object? Write a statement to generalize your observations.

4. Why is it important to include a number and a unit when reporting a measure?

Part B: Measuring Lengths with the Same Nonstandard Unit

Body lengths, such as the span of your hand, the length of your forearm, or the length of your foot, can also be used as non-standard units. Select one of these body parts as a unit and use it to measure three lengths in your classroom. Have two people measure the same lengths. Record their measures to the nearest whole unit in the given table and then answer the questions.

Object or distance	Measure of person 1	Measure of person 2

1. Describe how you used the units to obtain your measures.

2. a. How do the measures of the same objects compare? Can you explain any differences?
   b. What problems do the differences imply about using non-standard units?

Part C: Summary and Connections

1. Use what you have learned to answer the following questions.
    a. Bill used the length of his foot to measure the length of the hall and got 225 Bill feet. Mia used her foot to measure the length of the hall and got 210 Mia feet. What can you conclude from the measures?
    b. Polly used the length of a Popsicle stick to measure the length of a table. If she measured to the nearest whole Popsicle stick and got 24 sticks for the length, what can you conclude about the true length of the table?

2. What is the value in having students measure with nonstandard units before introducing standard units?

3. Search the Internet for the history of standardized units. Write a report on what you find.

Exploration 13.2 Understanding Metric Lengths

Purpose: Students use body parts to understand the relative size of metric lengths.

Materials: Centimeter ruler, meter stick, or metric tape measure

Part A: Metric Lengths on Your Body

Measures are easier to understand when we understand the relative size of the units used to make them. Developing a feel for the relative size of metric measures is difficult for many of us because we lack an intuitive sense of the length of metric units.

1. Work in pairs to measure each length in metric units. Be sure to record your measures.

    a. Width of your pinkie finger at the nail _____

    b. Length of your thumb _____

    c. Width of your closed hand _____

    d. Distance from your waist to the floor _____

    e. Length of your normal walking pace _____

2. Which body measure from Question 1 best approximates each of the following metric lengths? If none, find another length on your body that does.
    a. 1 cm   b. 1 dm   c. 1 m

Part B: Estimate Measures Using Metric Lengths

1. Use the information you gathered in Part A to estimate each length in metric units.
    a. Height of the door   b. Distance from the doorknob to the floor
    c. Length of a table   d. Width of your notebook
    e. Length of your cell phone   f. Length of the hall

2. Use a metric ruler or meter stick to check your estimates. How close were you?

Part C: Summary and Connections

1. Why is it helpful to use your body parts as a reference for understanding the relative size of metric length units?

2. If someone said that using body parts is not helpful because not everyone's body is the same size, how would you respond?

Exploration 13.3 Conversions Between English and Metric Lengths

Purpose: Students measure lengths to build a connection between English length and metric length units.

Materials: Ruler, tape measure, meter stick, tape, paper and pencil

Part A: Comparing Inches to Centimeters

Use a ruler to draw seven segments that are exactly 1, 2, 3, 4, 5, 6, and 7 inches in length. After drawing the segments, measure their lengths in centimeters. Record the measures in the table and answer the given questions.

Length in inches	1 in.	2 in.	3 in.	4 in.	5 in.	6 in.	7 in.
Length in centimeters							

1. Describe the relationship between a segment's length in inches and its length in centimeters.

2. Does the relationship exhibit a function? If so, what kind of function?

3. Make a graph by plotting the values in the table on the Cartesian plane.
    a. What does the graph reveal about the relationship between the two sets of measures?
    b. How does the graph confirm your answer to Question 2?

4. Write an equation that represents the relationship between the two sets of lengths.
    a. How could you use this equation to make conversions from inches to centimeters?
    b. How could you adapt your equation to make conversions from centimeters to inches?

5. Use your equation from question 4 to make each conversion.
    a. 45 in. to cm
    b. 3.67 in. to cm
    c. 14.81 in. to cm
    d. 35 cm to in.
    e. 24.6 cm to in.
    f. 1.39 cm to in.

Part B: Comparing Feet to Meters

Use a tape measure and tape to mark off five segments on your classroom floor that are exactly 3, 6, 9, 12, and 15 feet long. After marking the segments, measure their lengths in meters. Record your measures in the table and answer the given questions.

Length in feet	3 ft	6 ft	9 ft	12 ft	15 ft
Length in meters					

1. Describe the relationship between the length of a segment in feet and its length in meters.

2. Does the relationship exhibit a function? If so, what kind of function?

3. Make a graph by plotting the values in the table on the Cartesian plane.
    a. What does the graph reveal about the relationship between the two sets of measures?
    b. How does the graph confirm your answer to Question 2?

4. Write an equation that represents the relationship between the two sets of lengths.
    a. How could you use this equation to make conversions from feet to meters?
    b. How could you adapt the equation to make conversions from meters to feet?

5. Use the fact that 1 yd is equivalent to 3 ft to describe a relationship between the length of a segment in yards and its length in meters.
    a. Does the relationship exhibit a function? If so, what kind?
    b. Write an equation that represents the relationship between the length of a segment in yards and its length in meters.

6. Use your equations from Questions 4 and 5 to make each conversion.
    a. 8 ft to m
    b. 19.3 ft to m
    c. 6 m to ft
    d. 12.5 m to ft
    e. 19 yd to m
    f. 0.5 yd to m

Part C: Summary and Connections

1. a. Use what you have learned to create a conversion equation from
        i. yards to decimeters
        ii. feet to decimeters
        iii. centimeters to feet
        iv. inches to meters
    b. Describe the process you used to create the equations.

2. Will conversion equations from one length unit to another always be linear? Explain.

Exploration 13.4 Estimating Distances

Purpose: Students estimate distances using English and metric units.

Materials: String, rulers, and tape measures in both English and metric units

Part A: Estimating Lengths and Distances along a Line

The ability to estimate a length or distance is a useful skill, yet many people seldom practice it. The following questions give you that opportunity.

1. Estimate each length or distance in English and metric units. Record your estimates.
   a. Length of your writing utensil
   b. Length of your desk or table
   c. Width of your classroom
   d. Height of your classroom building
   e. Distance from your classroom building to another nearby building
   f. Distance from your classroom building to the nearest gas station

2. How did you decide which English unit to use? Which metric unit to use?

3. What methods did you use to make your estimates? Why did you choose one method over another to make each estimate?

4. Compare your estimates to those of some of your peers. How do they compare? What might account for any differences?

5. How could you check the accuracy of your estimates? Do so for those that are convenient.

Part B: Estimating the Length of a Curve

In Part A, the lengths and distances were along a line, making it relatively easy to make an estimate. Making an estimate can be more difficult if the length is around a curve.

1. Use English units to estimate the length of each curve. Record your estimates.
   a.
   b.

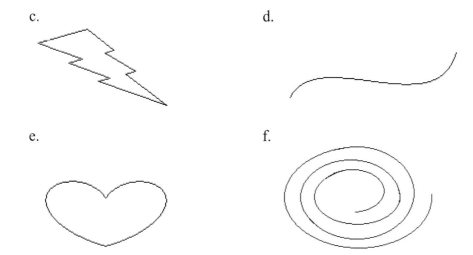

c.

d.

e.

f.

2. a. What methods did you use to make your estimates? Why did you choose one method over another to make each estimate?
   b. How were your methods for estimating the length of a curve similar to or different from the methods you used to estimate straight lengths and distances?

3. Compare your estimates to those of some of your peers. How do they compare? What might account for any differences?

4. How could you check the accuracy of your estimates? Do so for those that are convenient.

5. Select three or four curves from your surroundings. Estimate the length of each one, then check your estimate through direct measurement if possible.

Part C: Summary and Connections

1. In what situations in your life have you had to estimate a length or distance? What methods did you use to make your estimates?

2. a. Which do you think you do more often: estimate lengths or measure them directly? Why do you think this is the case?
   b. How might your answer to the previous question influence how you teach measurement in your classroom?

Exploration 13.5 Measuring Perimeters with the String Technique

Purpose: Students use string to make sense of perimeter as a measure of length.

Materials: Circular objects, scissors, string, metric ruler, a calculator, and paper

Part A: Measuring Perimeters

The **perimeter** of a simple, closed curve is defined to be the distance around the curve. For many curves, such as polygons, finding the perimeter is a matter of measuring segments and adding up their lengths. However, for others, the only way to directly measure the perimeter is to use the "string" method.

Draw several simple, closed curves on paper and cut them out. Be sure to include at least one or two that are not polygons. Measure the perimeter of each shape by taking a piece of string and wrapping it around the outside of the shape to find the point at which the string touches itself. After you find this point, straighten the string and measure its length with a metric ruler. Record the perimeter on the shape.

1. How does pulling the string straight show that a perimeter is a measure of length?

2. Measuring the perimeter of a polygon is relatively easy because the sides are segments that can be measured directly with a ruler.
    a. How could you use line segments to estimate the perimeter of any curve?
    b. Once you have an estimate, what could you do to the segments to make the estimate better?

Part B: Measuring Circumferences

The string method can used to find a special relationship that occurs in circles. Find three or four circular objects and use the string method to measure the circumference of each one in centimeters. Next, measure the diameter in centimeters. Record your measures in the table.

Circle	1	2	3	4
Circumference (C)				
Diameter (d)				

1. Compare the circumference of each circle to its diameter.
    a. Which is longer, the circumference or the diameter?
    b. Write a conjecture about any relationship you see between a circle's circumference and its diameter.

2. If you have not done so already, compute the ratio of the circumference to the diameter, or $\frac{C}{d}$, for each circle.

   a. Do the results support the conjecture you made in Question 1?
   b. Take the average of the ratios and compare the results to those of your peers. What do you notice?
   c. Write a generalization that expresses the relationship between the diameter of a circle and its circumference.
   d. Write a generalization that expresses the relationship between the radius of a circle and its circumference.

Part C: Summary and Connections

1. What problems did you encounter when using the string method to measure perimeters and circumferences? How do you think the problems affected your outcomes?

2. The ratio of the circumference to the diameter of any circle is always a constant called pi. Search the Internet for this number. Write a paragraph that summarizes what you find.

Exploration 13.6 Perimeter and Area—Are they Related?

Purpose: Students use their foot to explore the independence of perimeter and area.

Materials: Cardstock, dried beans that are approximately the same size, string, quarter-inch grid paper, square tiles, scissors, and tape

Part A: A Foot Problem

Trace your foot on cardstock and cut it out. Use your tracing to answer each question.

1. Estimate the perimeter of your foot in inches and write it down. Next, devise a method for directly measuring the perimeter of your foot in inches. Describe your method.
    a. What measure did you get for the perimeter of your foot?
    b. How did your measure compare to your estimate?

2. Using a bean as an area unit, estimate the area of your foot and write it down. Next, directly measure the area of your foot with beans.
    a. What measure did you get for the area of your foot?
    b. How did your measure compare to your estimate?

3. Estimate the area of your foot in square inches and write it down. Next, devise a method for directly measuring the area of your foot in square inches. Describe your method.
    a. What measure did you get for the area of your foot?
    b. How did your measure compare to your estimate?

4. Draw a square that has the same perimeter as your foot and measure its area.
    a. How does the area of the square compare to the area of your foot?
    b. Is there a relationship between the areas of two shapes with the same perimeter?

5. Make two new feet that are similar to yours; one with a perimeter that is half as long and one with a perimeter that is twice as long. Measure the areas of the new feet.
    a. How does the area of the smaller foot compare to the area of your foot?
    b. How does the area of the larger foot compare to the area of your foot?
    c. Do you see a relationship between the areas of two shapes that are similar?

Part B: Perimeters of Shapes with the Same Area

1. Use a set of 12 square tiles to make as many shapes as possible. Any two tiles that touch must do so completely along a side. Sketch your arrangements on grid paper and find the perimeter and area of each.

a. How do you know that each shape has the same area?

b. Do you see any patterns that suggest a relationship between the perimeter of a shape and its area?

2. Cut out a square that measures 5 in. × 5 in. Cut the square into any two pieces and put the pieces together to make several new shapes. The pieces should not overlap and must touch at edges and not corners. Record each shape and measure its perimeter.

a. How do you know that each shape has the same area?

b. Do you see any patterns that suggest a relationship between the perimeter of a shape and its area?

Part C: Areas of Shapes with the Same Perimeter

Cut a piece of string 20 in. long and tie the ends together to form a loop. Make a shape with your loop and measure its area by filling it with square tiles. Tape may be useful to hold the shape in place. Make a sketch of the shape and record its area. Repeat for three or four more shapes, making them as different from one another as you can.

1. How do you know that each shape has the same perimeter?

2. Do you see any patterns that suggest a relationship between the area of a shape and its perimeter?

3. Examine the shapes that held the most and least numbers of tiles. What observations can you make about the perimeters and areas of these shapes?

Part D: Summary and Connections

1. Summarize what you have learned about the relationship between perimeter and area. Can you determine the area of a shape from its perimeter? What about the perimeter of a shape from its area?

2. Perimeter and area are said to be independent concepts. What does this statement mean?

3. Suppose you want to build a rectangular pen for your dog. You have 24 meters of fencing in one-meter lengths. What are the dimensions of the possible rectangular pens? Are any of the pens more reasonable to use than others? Why?

Exploration 13.7 Misconceptions about Perimeter and Area

Purpose: Students investigate common misconceptions children have when learning about perimeter and area.

Materials: Paper and pencil

Part A: Identifying Misconceptions

In the classroom, many children have trouble discerning the difference between perimeter and area. As their teacher, it will be up to you to identify and clarify their misconceptions. Examine each scenario, identify the mistake, and explain how you might correct the error.

1. After counting, Monique claims that the perimeter of the following rectangle is 12 cm.

2. Deidre says, "If you cut a rectangle in half, the perimeter will be half its original length." How would you respond to Deidre?

3. Petra states that if a square and a rectangle have the same perimeter, the rectangle will always have the greatest area.

4. After cutting a rectangle into two pieces and rearranging them as shown, Hunter claims that the shapes have the same perimeter and area because they are made from the same pieces.

    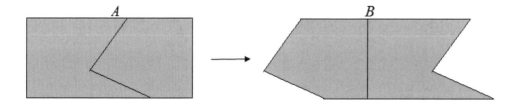

5. Matthew claims that the area of a shape is always bigger than its perimeter because area involves multiplication and perimeter is just adding lengths.

6. Taylor cuts a piece out of square *C* and throws it away so that only piece *D* remains. She then claims that the perimeter and area of shape *D* must be smaller than the perimeter and area of square *C* because a piece has been removed.

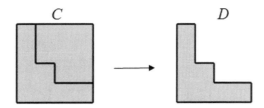

7. Abby claims that when comparing two rectangles, the one with the greater area will also have the greater perimeter.

Part B: Summary and Connections

1. How would you help a student make sense of the following problem?
   The area of square *A* is 16 cm², the side length of square *B* is 16 cm, and the perimeter of square *C* is 16 cm. Put squares *A*, *B*, and *C* in order of size, starting with the smallest.

2. What misconceptions, if any, have you had about perimeter and area?

Exploration 13.8 Pick's Theorem

Purpose: Students use a geoboard to discover Pick's Theorem.

Materials: 6 × 6 rectangular geoboard, and rubber bands.

Part A: Discovering Pick's Theorem

Georg Pick (1859 – 1942) is credited with a simple theorem that can be used to find the area of any polygon placed on a rectangular grid. It does so by relating the area of the shape to the number of grid points on the boundary and in the interior of the shape.

1. Create the following shapes on a 6 × 6 rectangular geoboard. Use the shapes to complete the table by finding the area of the shape (*A*), the number of pegs on the boundary of the shape (*b*), and the number of pegs in the interior of the shape (*i*). The area of each shape can be found by counting square units directly. A peg is on the boundary of the shape if the rubber band touches or goes around it.

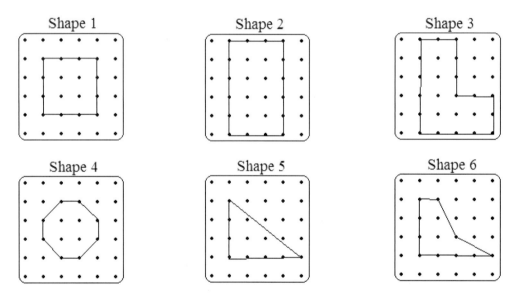

	Area (A)	Boundary pegs (b)	Interior pegs (i)
Shape 1			
Shape 2			
Shape 3			
Shape 4			
Shape 5			
Shape 6			

2. Use the table to look for Pick's theorem; that is, find a relationship that relates the area to the number of boundary and interior pegs. Express Pick's theorem with a formula in terms of A, b, and i.

Part B: Using Pick's Theorem

Pick's theorem can be used to find the area of concave polygons which do not easily lend themselves to standard area formulas. Make each shape on a 6 × 6 rectangular geoboard and then use Pick's theorem to find the shape's area.

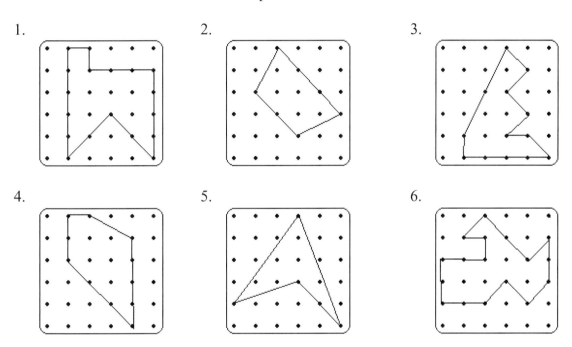

Part C: Summary and Connections

1. Why is it necessary for a polygon to be on a rectangular grid for Pick's theorem to work?

2. Do you think it is reasonable to use geoboards and Pick's theorem to demonstrate the concept of area in an elementary or middle grades classroom? Why or why not?

Exploration 13.9 Area Formulas of Polygons

Purpose: Students use transformations to find the area formulas for polygons.

Materials: Square-inch tiles, protractor, ruler, scissors, paper and pencil

Part A: The Area Formulas for a Rectangle and a Square

1. Use square tiles to make five rectangles with different lengths and widths. Complete the table by recording the length, the width, and the number of tiles needed to make each rectangle. After completing the table, describe the relationship between the length, the width, and the area. Use the variables, l, w, and A to write a formula that expresses the relationship.

	Length (l)	Width (w)	Number of Square Tiles
Rectangle 1			
Rectangle 2			
Rectangle 3			
Rectangle 4			
Rectangle 5			

2. A square is a rectangle in which the length and width are the same; that is, the length of each of the four sides is equal to s.
    a. Use your answer to Question 1 to describe the relationship between the area of a square and the length of its side.
    b. Use the variables s and A to write a formula that expresses the relationship.
    c. Use square tiles to create several different squares. Calculate the area of each square by using your formula. Check your answer by directly counting the number of square tiles.

Part B: The Area Formula for a Parallelogram

Use a ruler and a protractor to carefully construct a parallelogram that is not a rectangle. Construct it so that it has a base 5 in. long and a height of 3 in. The height of a parallelogram is the distance between the base and the opposite parallel side. Cut out the parallelogram.

1. Cover the parallelogram with square tiles to measure its area.
    a. Can you exactly cover the parallelogram with the tiles?
    b. Use what you see to make an estimate for the area of the parallelogram.

2. Use a protractor to draw a segment on the parallelogram that is perpendicular to the base. Cut the parallelogram along this segment and rearrange the pieces to make a rectangle.
   a. How does the area of the rectangle compare to the area of the parallelogram?
   b. Cover the rectangle with square tiles. What is its area?
   c. What is the area of the parallelogram? How does the area compare to your estimate?

3. If b represents the length of the base and h represents the height, use the area formula for a rectangle to write a formula for the area of a parallelogram in terms of b and h.

Part C: The Area Formula for a Triangle

To find the area formula for a triangle, the base and the height must be known. The base can be any of the three sides. The height is then the perpendicular distance from the vertex opposite the base to the line containing the base. Use a ruler to construct an acute triangle. Cut it out, make several copies on other sheets of paper, and then cut out the copies. Label the base and the height on all copies of the triangle.

1. Take two copies of the triangle and rearrange them to form a parallelogram.
   a. What is the base of the parallelogram? What is the height?
   b. How does the area of the parallelogram compare to the area of the triangle?
   c. Use the area formula for a parallelogram to write an area formula for a triangle in terms of the base, b, and the height, h.

2. Take a copy of the triangle and use a ruler to find the midpoints of the two sides that are not the base. Draw a segment connecting the midpoints. Cut the triangle along the segment and rearrange the pieces to form a parallelogram.
   a. Why must the new shape be a parallelogram?
   b. What is the base of the new parallelogram? The height? How does the area of the parallelogram compare to the area of the triangle?
   c. Use the new parallelogram to write an area formula for a triangle in terms of its base, b, and its height, h. How does the formula compare to the one you wrote in Question 1?

3. Take a copy of the triangle and use a ruler to draw the height to a base. Again find the midpoints of the two sides that are not the base. Connect the midpoints with a segment and cut the triangle along the segment. Then cut the triangular piece along the segment that was part of the height in the original triangle. Rearrange the pieces to form a rectangle.
   a. Why must the new shape be a rectangle?
   b. What is the length of the rectangle? The width? How does the area of the rectangle compare to the area of the triangle?
   c. Use the rectangle to write an area formula for a triangle in terms of its base, b, and its height, h. How does the formula compare to the one you wrote in Question 1?

4. What other ways can you cut or rearrange copies of the triangle to form other shapes that can be used to derive a formula for the area of the triangle?

Part D: Area Formulas for a Trapezoid

To find the area formula for a trapezoid, the bases and the height must be known. The bases are the parallel sides and the height is the distance between them. Use a ruler to construct a trapezoid. Cut it out, make several copies on other sheets of paper, and then cut out the copies. Label the bases and height on all copies of the trapezoid.

1. Take two copies of the trapezoid and rearrange them to form a parallelogram.
    a. Why must the new shape be a parallelogram?
    b. What is the base of the parallelogram? What is the height?
    c. How does the area of the parallelogram compare to the area of the trapezoid?
    d. Use the area formula for a parallelogram to write an area formula for a trapezoid in terms of the bases, b_1 and b_2, and the height, h.

2. The legs of a trapezoid are the two non-parallel sides. Take a copy of the trapezoid and locate the midpoints of the legs. Draw a segment connecting the midpoints. Cut the trapezoid along the new segment and rearrange the pieces to form a parallelogram.
    a. Why must the new shape be a parallelogram?
    b. What is the base of the new parallelogram? What is its height? How does the area of the parallelogram compare to the area of the original trapezoid?
    c. Use the new parallelogram to write an area formula for a trapezoid. How does the formula compare to the one you wrote in Question 1?

3. Take a copy of the trapezoid and use a ruler to draw one of its diagonals, subdividing the trapezoid into two triangles.
    a. How does the area of the two triangles compare to the area of the trapezoid?
    b. What are the bases of the triangles? What are their heights?
    c. Use the area formula of a triangle to derive an area formula for a trapezoid. How does the formula compare to the one you wrote in Question 1?

4. What other ways can you cut or rearrange copies of the trapezoid to form other shapes that can be used to derive a formula for the area of the trapezoid?

Part E: Summary and Connections

1. Why is it reasonable to use the area of a rectangle to model whole-number multiplication?

2. What transformations did you use to complete the various parts of this exploration?

Exploration 13.10 Finding the Area of a Circle

Purpose: Students makes sense of the formula for the area of a circle.

Materials: Unlined white paper, colored paper, scissors, tape, a ruler, and a compass

Part A: Using Radius Squares to Find the Area of a Circle

Use a compass to draw a circle on a sheet of paper and mark its center. Use a ruler to draw a radius of the circle. Measure the radius. Construct and cut out four squares from colored paper so that each square has sides equal to the length of the radius.

1. Compare the area of one square to the area of the circle. Which is larger?

2. Find the number of squares needed to cover the circle. If necessary cut the squares to be precise as possible. What is the area of the circle in radius squares?

3. a. Let r represent the radius of the circle. What is the area of one square in terms of r?
   b. Can you make a conjecture about the area of your circle in terms of r?

4. In the previous question, you derived a formula for the area of a circle. Does your derivation constitute a proof of the formula for the area of a circle? Explain.

Part B: Using Sectors to Find the Area of a Circle

Use a compass to draw a circle on a sheet of white paper and cut it out. Fold the circle in half along a diameter. Continue to fold it in half until it becomes difficult to do so. Unfold the circle and cut out the sectors. You should have about 8 or 16 congruent sectors.

1. Arrange the sectors into a parallelogram-like shape. How does the area of the circle compare to the area of the new shape?

2. Express the area of the parallelogram-like shape in terms of the circle's radius and circumference. Is this area formula equivalent to the one you obtained in Part A? How do you know?

3. Does this derivation constitute a proof of the formula for the area of a circle? Explain.

Part C: Summary and Connections

1. a. Which of the two strategies for finding the area of a circle do you prefer?
   b. Which one do you think upper elementary students would best understand? Why?

2. How could you use fraction disks to do the exploration in Part B?

Exploration 13.11 Understanding and Measuring Surface Area

Purpose: Students make measurements of surface area in various ways.

Materials: A box, a cylinder, a pyramid, a cone, paper, scissors, tape, measuring tools, and isometric dot paper

Part A: Surface Area and Nets

Wrap the box, cylinder, pyramid, and cone with sheets of paper so that the paper completely covers the shape but does not overlap itself. If necessary, use scissors and tape to help place the paper. Answer each question after you are finished.

1. Surface area is the amount of plane needed to "cover" a three-dimensional shape.
   a. Why is wrapping a shape in paper a good representation of surface area?
   b. Why does it make sense to measure surface area with square units?

2. Estimate the surface area of each shape. What strategies did you use to make your estimates?

3. Unwrap the paper covering each shape. The resulting figure made by the cuts in the paper is called a **net**. Describe the shapes that have been made by the folds and cuts in each net. Do the shapes suggest a way to compute the surface area of each shape? Explain.

4. Use a ruler or other measurement tool to find the surface area of each shape.
   a. What method did you use to obtain each measure? How did you use the net?
   b. Was your estimate close to the actual measure?

Part B: Measuring the Surface Area

It is also possible to find the surface area of less conventional shapes. Consider the following questions.

1. Find the surface area of each shape if one square is equal to one unit of area.
   a. 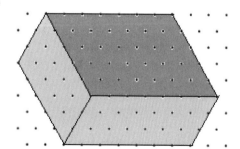   b.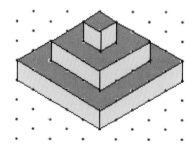

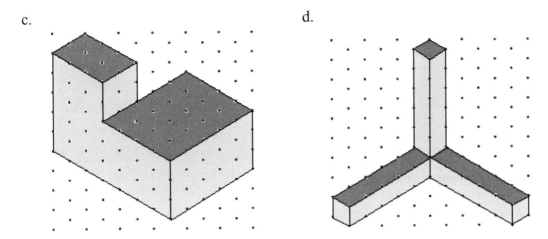

c.

d.

2. Use isometric dot paper to create several figures like the ones in Question 1. Switch drawings with a partner and find the surface area of each figure you are given.

3. What strategies did you use to find the surface area? How could you adapt your strategies if the shapes were not on isometric dot paper?

Part C: Summary and Connections

1. In Part A, you wrapped shapes in paper to find their surface area. Do you think this technique will work for any three-dimensional shape? Why or why not?

2. Why is it important to know area formulas when finding the surface area of a shape?

3. Spatial reasoning plays an important role in this exploration. Describe how you used it to complete Parts A and B?

Exploration 13.12 Understanding and Measuring Volume

Purpose: Students develop meaningful ways to measure the volumes of solids.

Materials: 1-in. grid paper, 1-in. cubes, scissors, tape, fillable geometric solids, rice, a funnel, and a pan or tray

Part A: The Volume of a Rectangular Prism

Draw the following nets on 1 in. grid paper. Cut them out, fold up the sides, and use tape to make open-topped boxes.

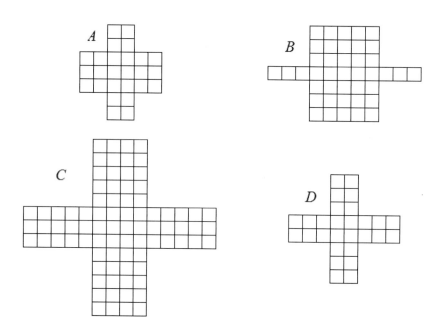

1. Set a 1-in. cube next to each box and estimate the number of cubes it will take to fill each box. Describe the strategy you used to obtain your estimates.

2. Use 1-in. cubes to completely fill each box. Count the number cubes it takes in each case. How do your measures compare to your estimates?

3. The unit of measure is a 1-in. cube or 1 cubic inch. Why does the unit of measure need to be three dimensional?

4. The open-topped boxes are examples of rectangular prisms. Based on your observations, how can you use measures of length to find the volume of any rectangular prism?

Part B: The Volume of Prisms and Cylinders

In Part A, the bases of the prisms were limited to rectangles. However, the base of a prism can be any polygon. Although this changes the shape of the prism, finding its volume is still a matter of finding the number of cubic units needed to fill it.

1. Consider the following triangular prism. How could you figure out how many cubes fit in the prism? What would you need to know?

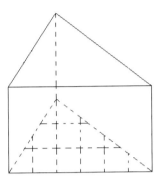

2. Would your strategy be different if the prism's base was a trapezoid? How would you find the volume of the following trapezoidal prism?

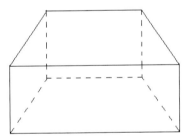

3. What would you do to find the volume of a cylinder? Is the strategy the same as finding the volume of a prism? Why?

4. In general, how can you find the volume of any right prism or cylinder? How might you represent this with a formula?

Part C: The Volume of Cones and Pyramids

1. Choose a cone and a cylinder from a set of fillable geometric solids so that they have the same base and height. Compare the volume of the cone to the volume of the cylinder by filling the solids with rice. What do you observe?

2. Choose a pyramid and a prism that have the same base and height. Compare the volume of the pyramid to the volume of the prism by filling the solids with rice. What do you observe?

3. Consider the following pentagonal prism. How could you use the volume of a prism to calculate the volume of the pyramid?

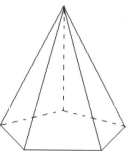

4. In general, how can you find the volume of any cone or pyramid? How might you represent this with a formula?

Part D: Summary and Connections

1. Explain the importance of the base and the height in calculating the volume of solids.

2. All the figures in this exploration were right, three-dimensional shapes. How could you compute the volume of the same shapes if they were oblique?

3. How could you use rice and an open-topped box with a volume of exactly 1 cubic inch to find the exact volume of any of your fillable geometric solids?

Exploration 13.13 Estimating the Volume of Ice Cream

Purpose: Students estimate volumes while eating ice cream.

Materials: A rectangular half-gallon of ice cream, sugar cones, ice cream scoop, and rulers

Part A: Using an Ice Cream Cone to Estimate Volume

Use a rectangular half gallon of ice cream and sugar cones to complete the following activity as a class.

1. Place the half gallon of ice cream where everyone can see. Have each student estimate the volume of the ice cream in cubic inches and cubic centimeters. Record the estimates.

2. Give each student a sugar cone and have them estimate its volume in both cubic inches and cubic centimeters. Record the estimates.

3. Have students measure the diameter and slant height of their sugar cones in inches and centimeters. Use the values to compute the volume of each ice cream cone in both units.
   a. How did the estimates compare to the computed values? Were the estimates better in inches or in centimeters?
   b. What other values had to be computed in order to find the volume of a cone?

4. Have students estimate the number of cones it will take to empty the half gallon of ice cream. Assume that each cone is filled, but not overflowing. Use the estimated number of cones to compute the volume of the ice cream in both cubic inches and cubic centimeters. How does this estimate compare to the estimates made in Question 1?

5. Test the estimates made in Question 4 by having students fill their cones. Again, each cone should be completely filled, but not overflowing. Students may eat the cones once they are filled. Continue to fill cones until the half gallon is empty. Record the number of cones it took to empty the half gallon.
   a. How does the actual number of cones compare to the estimates from Question 4?
   b. Use the actual number of cones and the volume of one ice cream cone to compute the volume of ice cream in the half gallon in cubic inches and cubic centimeters.

6. Look at the volume given on the ice cream package. Convert these values into cubic inches and cubic centimeters.
   a. How do the estimates from the previous questions compare to the actual value?
   b. Which method of estimation was most accurate?
   c. Is there a difference between the actual volume of ice cream and the volume computed using the sugar cones? What might account for any differences?

Part B: Summary and Connections

1. How would this exploration have been different if the carton of ice cream had been a cylinder rather than a rectangular prism?

2. In what situations do you make estimations of volume in your everyday life?

Exploration 13.14 The Volume of a Sphere

Purpose: Students find the volume of a sphere by relating it to the volume of a cylinder.

Materials: Modeling clay, overhead transparency sheets, scissors, and tape

Part A: Relating the Volume of a Sphere to the Volume of a Cylinder

Use modeling clay to make a sphere with a diameter between 3 cm and 5 cm. Make a cylinder with an open top and bottom by wrapping a transparency sheet around the sphere. The sphere should fit snugly inside the cylinder. Trim the transparency sheet so that the height of the cylinder is equal to the diameter of the sphere. Tape the cylinder so that it remains rigid.

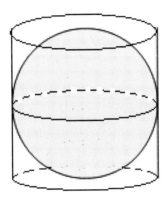

1. Suppose r represents the radius of the cylinder.
    a. Is r also the radius of the sphere? Explain.
    b. What is the height of the cylinder?
    c. Derive a formula for the volume of the cylinder in terms of r.

2. Flatten the sphere with your hand so that it fits snugly into the bottom of the cylinder. Be sure to fill the bottom of the cylinder completely. Mark the height of the flattened sphere on the cylinder. The height of the flattened sphere should be about $\frac{2}{3}$ the height of the cylinder.
    a. How does the volume of the flattened sphere compare to the volume of the cylinder?
    b. Use your answer to the previous question and the formula you developed in Question 1 to derive a formula for the volume of a sphere.

Part B: Apply What You Know

Use the formulas you derived in Part A to solve each problem.

1. a. What is the volume of a sphere with a radius of 4 in.?
    b. What is the volume of the smallest cylinder that would contain the sphere?

2. A silo is a building used to store grain. They often come in the shape of a cylinder with a hemisphere on top. What is the volume of a silo if it has a height of 100 ft and a radius of 20 ft?

3. a. Find the volume of a sphere with each radius.
      i. 3 in.   ii. 6 in.   iii. 12 in.
    b. How does the volume of the sphere with a 3-in. radius compare to the one with a 6-in. radius? How does the sphere with a 6-in. radius compare to the one with a 12-in radius?
    c. How does doubling the radius affect the volume of the sphere? Explain your thinking.

Part C: Summary and Connections

1. How does the volume of a sphere compare to the volume of a cone that has the same diameter as the sphere and a height equal to the diameter of the sphere?

2. Archimedes is credited with the discovery of the relationship between the volume of a sphere and the volume of the smallest cylinder that contains it. Search the Internet for other geometric discoveries found by Archimedes. Write a short report on what you find.

Exploration 13.15 Measuring Weight and Speed

Purpose: Students practice measuring weight and speed and consider the relationships between the units used to measure them.

Materials: Scales that measure in metric and English units, stop watch, meter sticks, and yard sticks

Part A: Measuring Weight

Weight is defined as the amount of force exerted by gravity on an object. In the sciences and in trade, metric units are commonly used to measure weights. However, many of us do not have an intuitive feel for the relative size of metric weights because we seldom use them in our daily lives. The following activities can help us make that connection.

1. Use a metric scale and an English scale to weigh at least four items in kilograms and in pounds. Record the items and the weights in the table.

Item	Weight in kg	Weight in lb

2. Graph each pair of weights on the Cartesian coordinate plane. Be sure to label your axes.
    a. What trends do you notice in the data?
    b. Why is the relationship between kilograms and pounds a function?
    c. What kind of function does it represent? How can you be sure?

3. Write an equation that represents the relationship between the weight of an object in kilograms and its weight in pounds. What does each number in the equation represent?

4. a. If an object weighs 14.5 kg, how much does it weigh in pounds?
   b. If an object weighs 85 lb, how much does it weigh in kilograms?

5. a. How many grams are in 1 kilogram?

   b. Use your answers to the previous questions to write an equation that represents the relationship between an object's weight in grams and its weight in pounds. Use it to find the weight, in grams, of each object in the table from Question 1.

Part B: Measuring Speed

Speed is the ratio of the distance traveled to the time of travel; that is, $\text{Speed} = \dfrac{\text{Distance}}{\text{Time}}$. It differs from other measures considered in this chapter because it is based on two other measures, a measure of distance and a measure of time. Measure and mark out a distance of 50 feet. Use a stop watch to measure the time it takes for you to walk the distance. Record the time.

1. Compute your walking speed in feet per second. What computations did you have to make?

2. What is your walking speed in miles per hour?

3. If there are 3.28 feet in one meter, what is your walking speed in meters per second? What about kilometers per hour?

4. Now walk the distance backward. Again record your time. Do you walk faster forward or backward? How much faster do you walk one way than the other?

Part C: Summary and Connections

1. Many people use the term weight as a synonym for mass, even though they represent different, yet connected, attributes. What is the difference between the two attributes? Search the Internet to find out.

2. In our everyday language, we often use the word "speed" to mean the rate at which a task is accomplished.
   a. In what situations have you used the word "speed" in this way?
   b. How does this use of the word "speed" compare and contrast with the scientific meaning given at the beginning of Part B?

Exploration 13.16 Measuring Temperature with Linear Functions

Purpose: Students use thermometers to collect data and discover a relationship between degrees Fahrenheit and degrees Celsius.

Materials: Celsius and Fahrenheit thermometers, various items with different temperatures

Part A: Comparing Celsius to Fahrenheit

Use a Celsius thermometer and a Fahrenheit thermometer to measure the temperature of at least four items with different temperatures. Record the information in the table.

Item	Celsius Temperature	Fahrenheit Temperature

1. Graph each pair of temperatures on the Cartesian coordinate plane. Be sure to label the axes. What trends do you see in the data?

2. a. Why is the relationship between degrees Fahrenheit and degrees Celsius a function?
   b. What kind of function does it represent? How can you be sure?

3. Write an equation that represents the relationship between degrees Fahrenheit and degrees Celsius. What does each number in the equation represent? Explain the meaning of the numbers in the context of the problem.

4. a. If the temperature is 75° F, what is the temperature in degrees Celsius?
   b. If the temperature is 25° C, what is the temperature in degrees Fahrenheit?

Part B: Summary and Connections

1. Do you think the relationship between the Celsius and Fahrenheit scales would be a good way to introduce functions to a class? Why or why not?

2. Why might a thermometer be a good tool for teaching integer concepts?

Chapter 14
Statistical Thinking

In the final two chapters, we turn away from geometry and measurement and begin a study of statistical thinking. **Statistical thinking** refers to the processes involved in collecting, organizing, and interpreting data. It is receiving more attention in the elementary grades for at least two reasons. First, by teaching statistical thinking at an early age, children have the opportunity to develop the information processing skills that will be essential to making decisions later in their lives. Second, statistical thinking provides an important way to build mathematical connections. It does so not only by providing a meaningful context in which students can apply concepts and skills from other areas of mathematics, but also by connecting mathematics to other disciplines and to students' daily lives.

Traditionally, statistical thinking in the elementary and middle grades has been limited to making and analyzing statistical graphs and computing averages. Modern curricula take a much more holistic approach. Now students are more likely to pose questions, design and conduct basic data collecting activities, represent data in different ways, and draw conclusions from the data to answer the questions they have posed.

The explorations in Chapter 14 are designed to take a hands-on approach to the different aspects of collecting and analyzing data. Specifically,

- Explorations 14.1 - 14.5 focus on posing questions, selecting samples, and collecting data in different ways.
- Explorations 14.6 - 14.9 use statistical graphs to represent and analyze data.
- Explorations 14.10 - 14.14 explore descriptive statistics.
- Exploration 14.15 considers misleading statistical graphs.

Exploration 14.1 Selecting Samples of Education Majors

Purpose: Students test simple random sampling by selecting samples for a small population in which the characteristics are known.

Materials: A table of random digits, the list of Current Education Majors

Part A: Gender of Education Majors

Simple random sampling is often used as the best way to select a sample that is representative of the population. To see if that is indeed the case, we can check randomly selected samples from a small population in which the characteristics are known.

For instance, consider the list of the current 100 education majors from a small university. Each student's identification number, gender, class, and age are given. Select 10 two-digit numbers from a table of random digits. Use the numbers as identification numbers to select 10 students from the list.

1. Count the number of males and females in your sample. What percent of your sample is male? Female?

2. Count the number of males and females in the population. What percent of the population is male? Female?

3. How do the percentages of males and females in your sample compare to those in the population? Were they close?

4. Compare the results of your sample with those of your classmates. In general, how do the percentages from the samples compare to the percentages in the population?

Part B: Class of Education Majors

Use a table of random digits and the identification numbers to select a new sample of 20 students.

1. Count the number students in each class: freshman, sophomores, juniors, and seniors. What percent of your sample is in each class?

2. Count the number of students in each class for the population. What percent of the population is in each class?

3. How do percentages of the students in each class in your sample compare to those in the population? Were they close?

4. Compare the results of your sample with those of your classmates. In general, how do the percentages from the samples compare to the percentages in the population?

Part C: Age of Education Majors

Use a table of random digits and the identification numbers to select a new sample of 25 students. Consider the number of traditional students to the number of non-traditional students. In this case, let a non-traditional student be any one over the age of 25.

1. Count the numbers of traditional and non-traditional students in your sample. What percent of your sample is traditional? Non-traditional?

2. Count the numbers of traditional and non-traditional students in the population. What percent of the population is traditional? Non-traditional?

3. How do the percentages of traditional and non-traditional students in your sample compare to those of the population? Were they close?

4. Compare the results of your sample with those of your classmates. In general, how do the percentages from the samples compare to the percentages in the population?

Part D: Summary and Connections

1. In general, were the characteristics of the samples representative of the characteristics of the population? What does this indicate about random sampling?

2. a. Which came closest to representing the population, the samples in Part A, Part B, or Part C?
   b. Why do you think there was a difference?
   c. Based on your answers to the previous questions, what can you do to make the sample more likely to be representative of the population?

3. What is the most difficult part of selecting a simple random sample?

Current Education Majors

ID#	Name	m/f	Class	Age	ID#	Name	m/f	Class	Age
01	Adams	F	Fr	18	51	Kirk,	F	Fr	18
02	Alexander	F	Jr	21	52	Klobuchar	M	Fr	21
03	Azzi	F	So	20	53	Kohl	F	So	20
04	Barrasso	F	Fr	19	54	Kyl	F	Jr	20
05	Baucus	F	Sr	34	55	Landrieu	M	Sr	29
06	Bell	F	Fr	18	56	Lautenberg	F	So	32
07	Bennet	F	So	19	57	Leahy	F	So	19
08	Bingamen	F	Jr	21	58	Lee	F	Jr	23
09	Blumenthal	F	Fr	19	59	Levin	F	Fr	25
10	Blunt	F	Sr	23	60	Lieberman	F	Fr	18
11	Booker	F	Fr	18	61	Lugar	F	Fr	19
12	Boxer	M	Jr	24	62	Manchin	F	Sr	21
13	Brown	F	So	20	63	McCain	F	So	20
14	Brown	F	Fr	19	64	McCaskill	M	Jr	27
15	Burr	F	Sr	22	65	McConnell	F	Fr	19
16	Cantrell	M	Jr	26	66	Menendez	F	So	28
17	Cardin	F	Fr	19	67	Merkley	F	Fr	31
18	Carper	F	So	20	68	Mikulski	M	Sr	23
19	Casey	F	Fr	20	69	Moran	F	Fr	19
20	Chambliss	F	Sr	23	70	Murkowski	M	Fr	19
21	Coats	F	Fr	19	71	Murray	M	Jr	20
22	Coburn	F	Fr	19	72	Nelson	F	So	28
23	Cochran	F	Jr	21	73	Nelson	F	Fr	19
24	Collins	M	So	19	74	Rand	F	Jr	22
25	Conrad	F	Sr	22	75	Portman	F	So	32
26	Coons	F	Fr	18	76	Pryor	F	Sr	28
27	Corker	F	So	20	77	Reed	F	Fr	18
28	Cornyn	F	So	21	78	Reid	F	Sr	22
29	Crapo	F	Fr	21	79	Risch	F	So	26
30	DeMint	F	Jr	41	80	Roberts	F	Jr	21
31	Durbin	F	Fr	19	81	Rockefeller	F	Fr	19
32	Enzi	F	So	19	82	Rubio	F	Sr	24
33	Feinstein	M	Jr	26	83	Sanders	F	So	23
34	Franken	F	Fr	18	84	Schumer	F	Fr	20
35	Gillibrand	M	Sr	23	85	Sessions	F	So	19
36	Graham	F	Fr	19	86	Shaheen	M	Fr	40
37	Grassley	F	Jr	21	87	Shelby	F	Jr	22
38	Hagan	M	So	20	88	Snowe	M	So	33
39	Harkin	F	Sr	25	89	Stabenow	M	Fr	18
40	Hatch	F	So	19	90	Tester	F	So	20
41	Heller	F	Sr	23	91	Thune	F	Fr	19
42	Hoeven	F	Fr	22	92	Toomey	F	Jr	23
43	Hutchinson	M	Sr	22	93	Udall	F	So	27
44	Inhofe	F	So	19	94	Udall	F	Jr	24
45	Inouye	F	Fr	18	95	Vitter	F	Fr	30
46	Isakson	F	So	21	96	Warner	F	So	19
47	Johanns	F	Jr	20	97	Webb	F	Fr	27
48	Johnson	F	Fr	19	98	Whitehouse	F	Jr	22
49	Johnson	F	So	20	99	Wicker	F	Fr	18
50	Kerry	F	Sr	24	100	Wyden	F	So	26

Exploration 14.2 Writing and Conducting a Survey

Purpose: Students write a survey and use it to collect data.

Materials: Paper and pencil

Part A: Writing a Survey and Collecting Data

Form a small group with two or three of your peers to collect data by writing and administering a survey. Choose a topic of personal interest and then write a survey of eight to ten questions to collect data about the topic. Try to use several different types of questions to reveal different points of view about your topic. As you prepare your survey, you may find it helpful to consider the following questions:

- Who is your target population? Are your questions relevant to this group?
- How will you select a sample from the population so that it is representative of the population?
- Do your questions get at the things you want to know? Are they too broad or too focused?
- Are there too many demographic questions (i.e. gender, ethnicity, etc.) that really have no connection to the topic of study?
- Should your questions provide possible selections or should they be more open-ended?
- Should the respondents answer anonymously so that you are more likely to get honest data?
- After the data is collected, how will it be analyzed?

After you are finished writing your survey, administer it to your target population. Each member of the group should collect data from five to ten people.

Part B: Analysis

After collecting your data, pool the results and analyze the data. Summarize your findings by preparing a poster to be shared with the class. Be sure to include the questions and results from your survey.

Part C: Summary and Connections

1. What was the most difficult part of writing and conducting your survey? Why?

2. If you were to do the survey again, what would you do differently?

Exploration 14.3 Collecting Data through Observations

Purpose: Students collect and analyze data by conducting an observational study.

Materials: Paper and pencil

Part A: Collecting Data through Observations

Form a small group with two or three of your peers to collect data through observations. Carefully choose a topic for your study so that the data can be collected through observations only. After selecting your topic, decide what observations need to be made to complete your study. As you prepare, you may find it helpful to consider the following questions:

- Who is your target population? Who or what will you observe to complete the study?
- How will you select a sample so that it is representative of the population?
- How will you make your observations?
- Is your study designed to make a comparison? If so, how will you make the comparison?
- After the data is collected, how will it be analyzed?

After you are finished with your preparation, collect data from your target population. Each member of the group should observe five to ten members of the population.

Part B: Analysis

After collecting your data, pool the results and analyze the data. Summarize your findings by preparing a poster to be shared with the class. Be sure to include a brief description of your topic and the results from your observations.

Part C: Summary and Connections

1. What was the most difficult part of preparing and conducting your study? Why?

2. If you were to do the study again, what would you do differently?

3. Which do you think is easier to conduct: an observational study or a survey? In what instances would a survey be more difficult? What about an observational study?

Exploration 14.4 The U.S. Census Bureau

Purpose: Students explore the website for the United States Census Bureau.

Materials: Internet access

Part A: The United States Census Bureau

Log on to a computer and go to the website of the United States Census Bureau (http://www.census.gov). Take a few moments to familiarize yourself with its contents. Once you feel comfortable with the site, use it to answer each question.

1. What are the main categories for data listed at the website?

2. What are the projected populations for each state in the given year?
   a. Hawaii in 2015
   b. California in 2015
   c. Georgia in 2020
   d. New York in 2020
   e. Florida in 2025
   f. Ohio in 2030

3. What is the state median income (3-year) for a family of
   a. 3 in Alabama?
   b. 4 in Iowa?
   c. 5 in New Mexico?

4. a. What was the latest adjusted annual rate for sales of new one-family houses?
   b. What were the latest total U.S. business sales?
   c. What was the rental vacancy rate in the second quarter of the past year?
   d. What was the home ownership rate in the second quarter of the past year?

5. Use the Facts for Features to answer the following questions for the past year.
   a. About how many trick-or-treaters were there on Halloween?
   b. How much was spent at family clothing stores for back-to-school shopping?
   c. What was the estimated U.S. population on July 4th?
   d. What percentage of women age 40 to 44 were mothers?
   e. What was the median annual earnings of women 15 or older who worked full time?

6. Continue to explore the website. Make a list of other facts or information that you discover or find interesting.

Part B: Summary and Connections

1. What problems did you have in navigating the website?

2. a. What resources does the website provide for teachers?
   b. Which ones might you use in your classroom? How would you do so?

| Exploration 14.5 Data Collection in the Classroom |

Purpose: Students use state curriculum standards for data analysis to design data collection activities for students in the elementary grades.

Materials: State curriculum standards for data analysis

Part A: Collecting Data through a Survey

Form a small group with two or three of your peers and take a few moments to familiarize yourself with your state's standards for data analysis in Grades 3 through 5. When you feel comfortable with the standards, design an activity that is appropriate for Grade 3 and for which students collect and analyze data using a survey. Once you are finished, discuss the following questions.

1. a. What is an appropriate number of questions to include on the survey?
   b. What are the advantages and disadvantages of having more or fewer questions?

2. What type of variables are children most likely to ask questions about? Why do you think this is the case?

3. How is the data from the survey most likely to be collected and represented?

4. What kind of questions could you ask to help your students analyze the data?

5. What modifications can you make so that your activity it is appropriate for Grade 4? Grade 5?

Part B: Collecting Data through an Observation

Repeat Part A, but this time, design an activity that collects data using observations. Answer the same questions and then discuss how data collection with a survey differs from data collection with observations.

Part C: Summary and Connections

Review a set of curriculum materials for Grades 3 through 5 that are commonly used in your state. Use them to answer the following questions.

1. Do the data collection activities included in the curriculum materials satisfy the objectives in your state standards?

2. How do your activities compare or contrast with those in the curriculum materials?

Exploration 14.6 Representing a Data Set with Multiple Graphs

Purpose: Students compare and contrast different graphs for representing quantitative and qualitative data.

Materials: Ruler, protractor, compass, paper, and pencil

Part A: Representing Quantitative Data

A variety of graphs can be used to represent sets of quantitative data. Use the following set of scores from a 60-point sixth grade mathematics test to make a dot plot, a stem and leaf plot, and a histogram. Be sure to include all necessary labels. Use the graphs to answer the following questions.

45	48	51	54	46	37	39	37	41	50
55	60	58	42	42	50	35	37	46	58
28	56	56	45	48	39	47	42	50	55

1. a. What was the highest score on the test? The lowest score?
   b. What score(s) occurred most frequently?
   c. Are there any gaps or clusters in the data? If so, for what data values?
   d. Which of the three graphs could you use to answer these questions? Does this imply an advantage of one type of graph over another?

2. a. Which was the easiest graph to make? The hardest?
   b. How could you use the dot plot to help you make the stem and leaf plot?
   c. How could you use the stem and leaf plot to help you make the histogram?

3. If 100 students had taken the test, which would be the best graph to represent the data: a dot plot, a stem and leaf plot, or a histogram? What if 1,000 students had taken the test?

4. Write a summary of the advantages and disadvantages of using dot plots, stem and leaf plots, and histograms to represent quantitative data. Which do you think is best? Why?

Part B: Representing Qualitative Data

A variety of graphs can also be used to represent qualitative data. Use the following data on the favorite colors of a class of third graders to make a frequency table (include the relative frequency), a pictograph, a bar graph, and a circle graph. Be sure to include all necessary labels. Use the graphs to answer the following questions.

red	blue	green	blue	blue	green	red	orange	red	purple
pink	blue	red	red	green	blue	red	pink	blue	yellow
red	red	blue	green	pink	green	blue	yellow	pink	red

1. a. What is the favorite color of the class?
   b. What is the least favorite color of the class?
   c. Which of the four graphs could you use to answer these questions? Does this imply an advantage of one type of graph over another?

2. a. Which was the easiest graph to make? The hardest to make?
   b. How could you use the frequency table to make the pictograph and bar graph?
   c. How could you use the frequency table to make the circle graph?
   d. What do your answers imply about the importance of frequency tables in representing qualitative data?

3. If you were given the favorite colors of 100 students, which do you think would be the best graph to represent the data: a frequency table, a pictograph, a bar graph, or a circle graph? What if you were given the favorite colors of 1,000 students?

4. Write a summary of the advantages and disadvantages of using frequency tables, pictographs, bar graphs, and circle graphs to represent qualitative data. Which one do you think is best? Why?

Part C: Summary and Connections

1. What kind of graphs do you find the easiest to make? The most difficult to make? What makes them so?

2. a. Which graphs can be used to represent multiple sets of data?
   b. Of these, which do you think is the best choices for representing quantitative data? Qualitative data?

Exploration 14.7 Making Statistical Graphs with a Spreadsheet

Purpose: Students use a spreadsheet to make statistical graphs.

Materials: A spreadsheet with graphing capabilities

Part A: Making Bar Graphs

A **spreadsheet** is a large electronic table that has the ability to store, organize, and make computations on large sets of numbers. Most of them have graphing capabilities, which can allow us to make a variety of statistical graphs efficiently and accurately.

1. Open a spreadsheet and take a moment to familiarize yourself with its graphing features. When you are ready, enter the following data. It shows the number of cases of meat that a restaurant used over one month. Use the spreadsheet to make a bar graph of the data. Be sure to include all necessary labels.
    a. Describe the steps you took to make the graph. What features did you have to edit to complete the graph and include all necessary labels?
    b. Why is a bar graph a good choice for representing this data?

Type of meat	Cases used
Chick breasts	63
Fish fillets	17
Hamburgers	65
Pork chops	23
Steaks	48

2. Make a double bar graph of the following data. It shows the number of cases of meat that a restaurant used in two months. Be sure to include all necessary labels.
    a. What feature(s) did you have to include in the second graph that you did not have to include in the one from Question 1?
    b. How did the meat consumption change from January to June? Can you explain why the changes might have occurred?

Type of meat	Cases used in January	Cases used in June
Chick breasts	63	67
Fish fillets	17	30
Hamburgers	65	84
Pork chops	35	21
Steaks	48	23

Part B: Making Line Graphs

Spreadsheets can also make line graphs. To do so, you may need to use a scatterplot feature rather than an actual line graph.

1. Make a line graph of the following data. It shows the change in a school's population over a 40-year period. Be sure to include all necessary labels.
    a. Describe the steps you took to make the graph. What features did you have to edit to complete the graph and include all necessary labels?
    b. Why is a line graph the best choice for showing this data?
    c. What population changes does the graph reveal?
    d. What might account for the decrease in population between 1995 and 2000?

Year	1970	1975	1980	1985	1990	1995	2000	2005	2010
Number of Students	1,674	2,041	2,423	2,856	3,597	4,780	2,400	2,691	2,857

2. Make a line graph of the following data set. It shows the changes in the ethnicity of a school population over a 40-year period. Be sure to include all necessary labels. Once you are finished, describe how the ethnicity of the school has changed.

Year	1970	1975	1980	1985	1990	1995	2000	2005	2010
Percent African American	5.4%	3.3%	5.8%	6.2%	8.9%	15.3%	17.9%	20.4%	21.7%
Percent Caucasian	93.4%	95.6%	92.4%	91.5%	88.6%	82.1%	79.3%	75.6%	69.3%
Percent Hispanic	1.2%	1.1%	1.8%	2.3%	2.5%	2.6%	2.8%	4%	9%

Part C: Making a Circle Graph

Use a spreadsheet to make a circle graph of the following data set. It shows the percent of students that earned a particular letter grade on a final exam. Be sure to include all necessary labels. Why is a circle graph a good choice for representing this data?

Letter grade	A	B	C	D	F
Percent of students	8%	23%	47%	19%	3%

Part D: Summary and Connections

1. What, if anything, did you find difficult about using a spreadsheet to make statistical graphs? What features of the graphs did you most commonly have to edit?

2. Is it possible to use a spreadsheet to make a dot plot? What about a stem and leaf plot or a pictograph?

Exploration 14.8 Correlation and the Line of Best Fit

Purpose: Students use a spreadsheet to investigate the correlation between two variables and to find the line of best fit.

Materials: A spreadsheet with graphing capabilities

Part A: High School and College GPAs

The following table shows the high school and college grade point averages (GPA) for 24 students. The high school GPA is based on a 5.0 scale. The college GPA has been calculated after the freshman year of college and is based on a 4.0 scale. Use a spreadsheet to make a scatterplot of the data and answer the following questions.

H.S. GPA	College GPA	H.S. GPA	College GPA	H.S. GPA	College GPA
3.53	2.70	2.55	2.05	3.77	2.95
2.76	1.95	2.89	2.35	3.45	2.65
4.11	3.33	3.08	3.11	4.41	3.75
4.87	3.80	4.15	3.34	4.90	3.89
2.83	2.67	4.21	3.40	3.33	2.70
2.95	2.80	3.84	3.05	3.08	2.50
1.95	2.05	3.40	3.00	4.05	3.56
4.33	3.56	3.05	2.75	4.22	3.45

1. a. What are the two variables in the data set?
   b. Does the scatterplot show a positive or a negative correlation between them? Why might this relationship exist?

2. Use the spreadsheet to find the correlation coefficient. A **correlation coefficient** is a number that shows the strength of the relationship between the two variables. Numbers closer to 1 or -1 indicate a strong positive or negative correlation. Numbers closer to 0 indicate a weak correlation to no correlation.
   a. What steps did you take to find the correlation coefficient?
   b. Is the correlation coefficient positive or negative?
   c. Is there a strong or weak relationship between the variables?

3. Use the spreadsheet to find the regression line. A **regression line** is one line that best fits the linear pattern of the data.
   a. What steps did you take to find the regression line?
   b. Use the spreadsheet to find the equation of the regression line. Use the equation to find the likely college GPA of a student with a high school GPA of
      i. 2.25.   ii. 3.46.   iii. 3.98.   iv. 4.67.

4. Do you think a student's high school GPA is a good predictor of his or her GPA after the freshman year of college? Why or why not?

Part B: Hours of Exercise and Percent Body Fat

The following table compares the hours of exercise per week to the percent body fat for 20 women. Use a spreadsheet to make a scatterplot of the data.

Percent Body Fat	Hours of Exercise	Percent Body Fat	Hours of Exercise
33%	1	5%	14
7%	12	24%	1
12%	8	19%	2
10%	8	15%	4
10%	9	16%	3
21%	3	12%	5
14%	5	17%	3
15%	6	25%	2
11%	7	28%	1
13%	7	13%	6

1. a. What are the two variables in the data set?
   b. Does the scatterplot show a positive or negative correlation between them? Why might this relationship exist?

2. Use the spreadsheet to find the correlation coefficient.
   a. Is the correlation coefficient positive or negative?
   b. Is there a strong or weak relationship between the variables?

3. Use the spreadsheet to find the regression line and its equation. Use the equation to find the percent body fat of a woman who exercises for
   a. 3 hours.   b. 4.5 hours.   c. 9 hours.   d. 15 hours.

Part C: Summary and Connections

1. Revisit Part A. Find the ten lowest college GPAs and add 0.5 to each average. How does this affect the correlation coefficient and regression line?

2. Why is it useful to know when two variables are correlated?

3. Can you think of several pairs of variables that are relevant to the elementary classroom and are positively correlated? What about negatively correlated?

Exploration 14.9 Understanding Graphs with Percents

Purpose: Students examine two situations in which percents are used in graphs.

Materials: Paper and pencil

Part A: Pie Graphs and Percents

Students often have trouble making sense of graphs with percents because the percents only show relative amounts and not total values. For instance, consider the following two pie graphs which show how a monthly household income was spent in two different years. Use what you see in the graphs to answer the following questions.

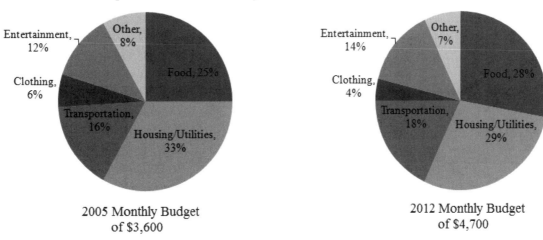

Expenditures of a Monthly Household Income for Two Years

2005 Monthly Budget of $3,600

2012 Monthly Budget of $4,700

1. In which year was the monthly income the largest?

2. In which year was a larger percentage of the budget spent on
   a. Food?   b. Housing/Utilities?   c. Transportation?   d. Clothing?

3. Calculate the actual amount of money spent on Housing/Utilities in both years. Why might the percents in the graphs be misleading?

4. Is there any expenditure for which less money was spent in 2012 than in 2005? Why do the percents make it possible to answer this question incorrectly?

Part B: Bar Graphs and Percents

The following bar graph shows the changes in enrollment at a new elementary school which opened in 2008. Use what you see in the graph to answer the following questions.

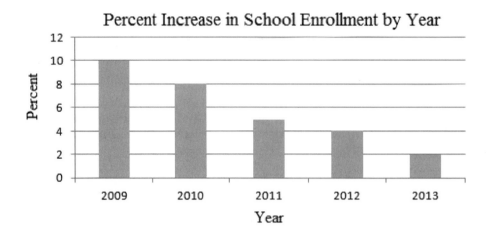

1. a. In what year was the percent increase the largest? The smallest?
   b. Are the percents increasing or decreasing?

2. Suppose that student enrollment in 2008 was 600 students. Find the number of students enrolled in each year.
   a. Did the number of students increase or decrease?
   b. Did the increase or decrease happen at a constant rate? Explain.

3. Why might the percents in the graph be misleading?

Part C: Summary and Connections

1. Search newspapers and magazines for graphs that contain percents. Write at least three questions that can be answered from the graphs.

2. a. What are possible errors that students might make when working with graphs with percents?
   b. What might make constructing a graph with percents difficult for students?

Exploration 14.10 Understanding Measures of Center

Purpose: Students investigate the mean, the median, and the mode.

Materials: A ruler, a pencil, pennies, and a spreadsheet

Part A: Understanding the Median

The **median** indicates the center of the data by giving the middle most number in a set of data that has been arranged in numerical order. Answer the following questions to get a better understanding of the median.

1. The following data set has 11 numbers arranged in ascending order. Find the median of the data set and describe the process you used to find it.

   14   15   17   20   20   24   25   26   26   27   29

2. The following data set has 10 numbers arranged in descending order. Find the median of the data set and describe the process you used to find it.

   20   23   27   29   33   35   38   38   40   45

3. a. Why is it important that the numbers be arranged in numerical order?
   b. Does it matter whether the numbers are arranged in ascending or descending order? Explain your thinking.
   c. Write a general procedure for finding the median of any data set.

4. a. What percent of the data is above the median? Below the median?
   b. Why do these percents make sense?

Part B: Understanding the Mean

The **mean** gives the center of the data by adding the data values then dividing by the number of data points. To get an intuitive understanding of the mean and its relationship to the data set, construct a simple balance by placing a ruler across a pencil or other writing utensil. The ruler should balance at the 6 in. mark. Place pennies on the ruler as directed and then answer each question.

1. Place a penny at the 3-in. mark and another at the 9-in. mark. Move the ruler to make it balance. How does the balancing point compare to the mean of 3 and 9?

2. Place a penny at the 2-in. mark and another at the 9-in. mark. Move the ruler to make it balance. How does the balancing point compare to the mean of the two numbers?

3. Place a penny at the 1-in. mark, another at the 7-in. mark, and another at the 10-in. mark. Move the ruler to make it balance. How does the balancing point compare to the mean of the three numbers?

4. Place the pencil underneath the 7-in. mark and place a penny at the 10-in. mark. How can you use the mean to predict where to place another penny so that the ruler balances? Place a penny at this location to test your prediction.

5. Intuitively, what does the mean represent for any data set?

Part C: Comparing the Mean, the Median, and the Mode

Enter the following data set of eleven numbers into a spreadsheet and use the spreadsheet to calculate the mean, the median, and the mode. Describe the procedure you used to do so.

    40   45   50   50   50   55   55   60   65   65   70

1. What are the values of the mean, the median, and the mode?

2. What happens to the mean, the median, and the mode if
    a. the 70 is changed to a 100?   b. the 40 is changed to a 20?

3. What happens to the mean, the median, and the mode if
    a. one 50 is a changed to a 55?
    b. one 55 is changed to a 50 and the other is changed to a 60?

4. Change numbers in the data set to match the following conditions. Record the set of numbers that satisfy each condition. Reset the numbers to their original values for each part.
    a. The mean, the median, and the mode are equal
    b. The mean is greater than the median and the median is greater than the mode
    c. The median is greater than the mean and the mode
    d. The mean and median are equal and smaller than the mode

5. Revisit the conditions for the data set given in Question 4. What generalizations can you make for any data set in which the given conditions hold?

Part D: Summary and Connections

1. Summarize what you have learned about the mean, the median, and the mode. Be sure to include how each measure is affected by changes in the data.

2. Which measure of center do you think is the most useful? Why?

Exploration 14.11 Understanding Box Plots

Purpose: Students interpret box plots and use them to compare data sets.

Materials: Yard stick or tape measure

Part A: Interpreting Box Plots

One way to show the spread of a data set is with a five-number summary, which gives the minimum, maximum, median, and upper and lower quartiles of the data. Five-number summaries are commonly shown with **box-and-whiskers plots,** or **box plots**. The plot contains an axis that shows the entire range of the data. The median, the upper quartile, and the lower quartile are shown with short line segments over the appropriate numbers on the axis. The segments are connected to form the box. The maximum and minimum are plotted as dots and then connected to the box with lines to form the whiskers. If there is an outlier, it is marked with an asterisk and the next highest or lowest value is connected to the box. For instance, the following box plot is for a set of scores on a 100-point exam. Use it to answer the following questions.

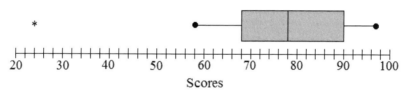

Scores on a 100-point Test

1. a. What are the maximum and minimum of the data?
   b. What is the range of the data?

2. Is there an outlier? If so, what is its value?

3. a. What is the median of the data? The upper quartile? The lower quartile?
   b. What percent of the data lies between the upper quartile and the median?
   c. What percent of the data lies between the median and the lower quartile?
   d. What percent of the data lies between the upper and lower quartiles?

4. Overall, how did the class do? Explain your thinking.

Part B: Comparing Box Plots

Box plots can also be used to compare data sets. Use the given box plot to answer each question.

386 Chapter 14 Statistical Thinking

1. The following box plots show the points scored per game by the players on two basketball teams.

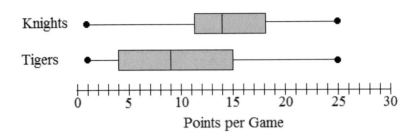

   a. How do the best shooters on the teams compare? The worst shooters?
   b. Which team has the greatest range of ability?
   c. If the two teams played, which one is most likely to win? Explain.

2. The following box plots represent the distribution of ages in months of 14 boys and 13 girls in a sixth grade class.

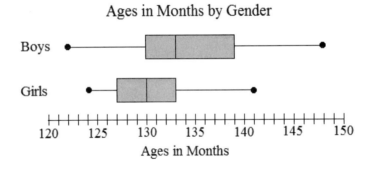

   a. Give five questions that can be answered by comparing the two box plots. Avoid asking the same question for each box plot—for example, avoid asking such questions as "What is the maximum age of the girls?" and "What is the maximum age of the boys?" Provide answers for your questions.
   b. Give three questions that cannot be answered by comparing the box plots.

Part C: Spanning a Distance

Divide the class into groups of 10 to 15 students as your instructor directs. Each group should use a yard stick or tape measure to make a segment exactly 9 ft long. Have each student in the group measure the distance by using his or her span. A **span** is the distance between the tip of the thumb and the tip of the pinky finger when the hand is outstretched. Record the data.

1. Have each group find the five-number summary for their data and represent it with a box plot. What do the box plots indicate about the data for each group?

2. Represent all box plots on a single axis so they can be compared.
    a. What comparison statements can you make based upon the box plots?
    b. Why is it possible to use the box plots to make a comparison between the groups even though the groups are not necessarily the same size?

Part D: Summary and Connections

1. Write a short summary of how five-number summaries and box plots show the spread of a set of data.

2. Is it possible to find the mean or the mode from a box plot? Why or why not?

3. Box plots are generally introduced in the middle grades. Why do you think they are not a part of the elementary curriculum?

Exploration 14.12 Understanding the Standard Deviation

Purpose: Students use a spreadsheet to investigate the relationship between the mean and the standard deviation.

Materials: A calculator and a spreadsheet

Part A: Computing a Standard Deviation

The **deviation** of any data point x is its distance from the mean, or $x - \bar{x}$. Intuitively, we can think of the **standard deviation** as the typical distance of all the data points from the mean; that is, the average of the deviations. Consequently, to compute a standard deviation, we compute every deviation and then take their average. However, we encounter a problem when we try to do so. To understand the problem and how we get around it, answer the following questions using the following data set:

$$6 \quad 8 \quad 8 \quad 10 \quad 11 \quad 11 \quad 12 \quad 14$$

1. a. What is the mean of the data?
   b. Why must we compute the mean to compute the standard deviation?

2. a. A deviation is always computed by subtracting the mean from the data point, or $x - \bar{x}$. Complete the table by computing the deviation for each data point.

x	$x - \bar{x}$
6	
8	
8	
10	
11	
11	
12	
14	

   b. Some deviations are positive and others are negative. In any data set, which values will have negative deviations and which will have positive deviations?

3. a. Compute the average of the deviations. What happens?
   b. Can you explain why the average of the deviations will always be equal to this value?
   c. What interpretation does this value have in terms of the typical distance of the data values from the mean?

4. The average of the deviations is zero because the net sum of the positive and negative deviations is zero. One way to handle the problem is to take the absolute value of the deviations before taking their average. This average is called the **mean absolute deviation**. Compute the mean absolute deviation for the given data. Interpret its value in terms of the typical distance of the data values from the mean.

5. In the middle grades, the mean absolute deviation is often used as an intermediate step to understanding the standard deviation. The standard deviation handles the problem of the negative deviations by squaring them first and then taking their average. The average of the squared deviations is called the **variance, v**.
   a. Complete the table by computing the squared deviation for each data point.

x	$x - \bar{x}$	$(x - \bar{x})^2$
6		
8		
8		
10		
11		
11		
12		
14		

   b. Compute the variance by taking the average of the squared deviations. How does the value of the variance compare to the values in the original data set?

6. The variance is the square of the needed value. To find the **standard deviation, s,** we take the square root of the variance.
   a. Compute the standard deviation for the given data set.
   b. Interpret the standard deviation in terms of the typical distance of the data values from the mean.
   c. How does the standard deviation compare to the mean absolute deviation? Which one do you think best represents the spread of the data and why?

Part B: Relating the Standard Deviation to the Mean

To understand the standard deviation as a measure of spread, it is necessary to understand its relationship to the mean. Enter the following data into a spreadsheet and use the spreadsheet to compute the mean and the standard deviation. Use what you see in the spreadsheet to answer the following questions.

```
35  36  42  47  48  48  50  50  53  54
56  57  60  61  62  63  65  68  71  73
```

1. a. What is the mean? Describe how the data is distributed around this value.
   b. What is the standard deviation? What does it tell you about the typical distance of all the data points from the mean?
   c. How many data points are within one standard deviation above or below the mean?

2. Change the 35 to a 20 and the 73 to a 90. The spreadsheet will automatically recalculate the mean and the standard deviation.
   a. How did changing the two values change the spread of the data?
   b. How did the mean change? The standard deviation change?
   c. What does the new standard deviation tell you about the typical distance of all the data points from the mean?

3. Change the three smallest numbers to 47s and the three largest numbers to 65s.
   a. How did changing the values change the spread of the data?
   b. How did the mean change? The standard deviation change?
   c. What does the new standard deviation tell you about the typical distance of all the data points from the mean?

4. Change the data back to their original values and then add 10 to each of the numbers.
   a. How did changing the values change the spread of the data?
   b. How did the mean change? The standard deviation change?
   c. What does the new standard deviation tell you about the typical distance of all the data points from the mean?

5. Change all of the numbers to 60.
   a. How did changing the values change the spread of the data?
   b. How did the mean change? The standard deviation change?
   c. What does the new standard deviation tell you about the typical distance of all the data points from the mean?

6. Write a general statement that relates the spread of the data to the size of the standard deviation.

Part C: Summary and Connections

1. Do you find it easy or difficult to understand the standard deviation as a measure of spread? Why?

2. When you computed the variance, you divided the sum of the squared deviations by n, the number of data points. It is more common to divide the sum of the squared deviations by $n - 1$. Search the Internet to find out why.

Exploration 14.13 Understanding the Normal Distribution

Purpose: Students explore a normal distribution to understand the standard deviation and its relationship to the mean.

Materials: Paper and pencil

Part A: Standard Deviation and a Distribution

A **distribution** is a graph that compares data values to the relative frequency of those values. For small data sets, a distribution is often represented with a dot plot. The following dot plot shows a distribution of 90 data values that have a mean of $\bar{x} = 59.2$ and a standard deviation of $s = 6.4$. Use the distribution, its mean, and its standard deviation to answer each question.

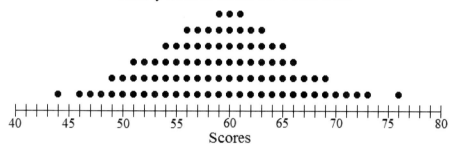

1. a. What value is one standard deviation above the mean; that is, equal to $\bar{x} + s$?
   b. What value is one standard deviation below the mean; that is, equal to $\bar{x} - s$?
   c. What percent of data values are between the mean and one standard deviation
      i. above the mean?   ii. below the mean?
   d. About what percent of the data lies between one standard deviation below and one standard deviation above the mean?

2. a. What value is two standard deviations above the mean? Two standard deviations below the mean?
   b. What percent of data values are between the mean and two standard deviations
      i. above the mean?   ii. below the mean?
   c. About what percent of the data lies between two standard deviations below and two standard deviations above the mean?

3. Revisit your answers to the last two questions. Summarize how the standard deviation can be used to describe the spread of the data relative to the mean.

Part B: Normal Distributions and the Empirical Rule

For larger data sets, a **normal distribution** is characterized by a smooth bell-shaped curve that theoretically extends to infinity without touching the horizontal axis. If a distribution is normal, then we know what percent of the data falls within a certain distance of the mean. Specifically, about 68% of the data lies within one standard deviation of the mean, 95% of the data lies within two, and 99.7% of the data lies within three. This fact is called the **empirical rule.** The following picture shows specific percentages within a normal distribution.

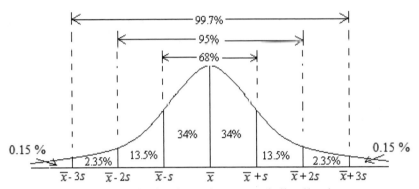

The empirical rule and a normal distribution

1. How do your percentages from Part A compare to the percentages in the empirical rule? What might account for any discrepancies?

2. Suppose a standardized test with a mean of $\bar{x} = 125$ and a standard deviation of $s = 10$ has scores that are normally distributed.
    a. Draw a picture of the distribution similar to the one above and label the values that are one, two, or three standard deviations from the mean.
    b. What percent of test takers had a score between
        i. 115 and 125?   ii. 125 and 135?   iii. 115 and 135?
    c. What percent of test takers had a score between
        i. 105 and 125?   ii. 125 and 145?   iii. 105 and 145?
    d. What percent of test takers had a score
        i. between 115 and 145?   ii. above 135?   iii. below 105?

3. The heights of 200 fourth graders are recorded and are found to be approximately normal. If the mean is 50 in. and the standard deviation is 3 in., then 95% of the students are between what two heights?

Part C: Summary and Connections

1. How does the empirical rule help you make sense of the standard deviation and its relationship to the mean?

2. Do you think the empirical rule holds for distributions that are not normal? Why or why not?

Exploration 14.14 Percentiles

Purpose: Students investigate percentile scores.

Materials: Paper and pencil

Part A: Understanding Percentiles

Percentiles are often used in education to compare the academic achievement of students. For any set of numbers, the ***n*th percentile** is a number P_n such that n percent of the numbers are less than or equal to P. A percentile indicates the position of a number relative to the other numbers in the data set. For instance, if a student scores at the 80th percentile (P_{80}), then 80% of the students who took the test made the same or a lower score. We can also say that 20% of the students had a higher score.

1. What percentile separates the lower 12% of values from the upper 88% of values?

2. The percentile P_{87} separates what upper percent of values from what lower percent of values?

3. What percentile is the cutoff point for the upper 35% of the values?

4. What is the cutoff point for the lower percentage of values represented by P_{35}?

5. If a student scores at the 80th percentile on a test, does that mean the student got 80% of the test questions correct? Explain.

Part B: Problem Solving with Percentiles

Percentiles are often useful when considering large data sets. For instance, suppose a university gives a comprehensive final exam to all college algebra students. Because there is a 10-point curve on the exam, scores can range from 10 to 110. Use percentiles to answer the following questions. Every percentile has a value between 1 and 99.

1. Suppose 20% of the students who took the exam made a grade of 40 or below.
    a. If Bill made a 40 on the final exam, at what percentile did he score?
    b. What percent of the students made a grade higher than Bill?
    c. Why is it impossible to tell how many students made a grade higher than Bill?

2. Suppose the highest grade on the final exam was a grade of 90. If Ramona made a 90 on the final exam, at what percentile did she score?

3. The scores of all students that took the college algebra final exam in Fall 2012 are shown in the table.

Score	30	35	40	45	50	55	60	65	70	75	80	85	90	95	100	105	110
Frequency	15	12	20	48	63	70	58	73	75	69	55	52	43	27	18	4	1

    a. What score is at the
        i. 99th percentile?   ii. 1st percentile?   iii. 35th percentile?
    b. What percentile is a score of
        i. 50?   ii. 70?   iii. 95?
    c. How many students scored at the 70th percentile or above?
    d. How many students scored at the 40th percentile or below?

Part C: Summary and Connections

1. What are other names for the 25th, 50th, and 75th percentiles of a data set?

2. Search the Internet for uses of percentiles in education. Write a short report on what you find.

Exploration 14.15 Misleading Statistical Graphs

Purpose: Students identify ways that data are misrepresented in statistical graphs.

Materials: Internet access, newspapers, or magazines, paper and pencil

Part A: Identifying Misleading Graphs

Identify how each graph misrepresents the data. Suggest how the graph can be changed so that it is no longer misleading.

1.

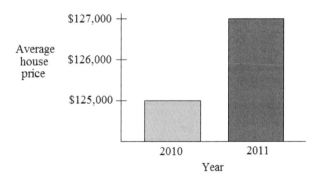

2.

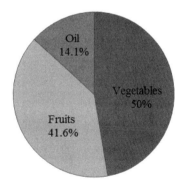

3.

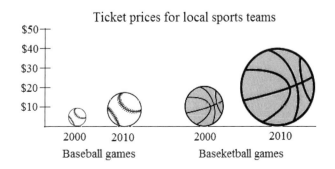

Part B: Making Misleading Graphs

1. The following table shows the profits earned by a small company in millions of dollars. Draw two line graphs of the data, one in which the vertical axis starts at 0 and another that starts at 12 million. How does the change in the vertical axis affect the perception of the graph?

Year	2006	2007	2008	2009	2010	2011	2012
Profits earned in millions of dollars	12.4	12.9	13.2	13.8	14.7	14.9	15.1

2. On weekdays, a convenience store sells an average of 850 cups of coffee a day. On weekends, the same store only sells an average of 420 cups of coffee a day. Draw a graph that uses the area or volume of a figure to distort this information.

Part C: Summary and Connections

1. Search the Internet, newspapers, or magazines for statistical graphs that are misleading.
    a. Describe how the graphs are misleading and how you could change them so they are no longer misleading.
    b. Write a summary of the ways you find that statistical graphs can be misleading.

2. Do you think it is important for students to learn about misleading graphs? Why?

Chapter 15
Probability

Probability is a topic closely related to statistics and data analysis, but it is also an important topic of study in its own right. It provides an important link between mathematical topics such as number, proportional reasoning, algebra, and geometry. It also provides a valuable context for solving problems, reasoning mathematically, developing number sense, and applying computational skills. Because of its practicality and importance, it is becoming more prevalent in the classroom.

Although children in the early elementary grades do not compute many probabilities, they do develop an intuitive notion of it using words such as *likely*, *certain*, and *impossible*. After they have a working knowledge of fractions, students can begin to explore probability as a measure of chance by conducting simple experiments, such as rolling dice or spinning spinners. In the middle grades, students learn that there are two types of probability: experimental and theoretical. Students will naturally encounter the need for experimental probability as they look to make predictions from data. They will also learn to use theoretical probability as they conduct probabilistic experiments using a variety of tools and simulations.

The explorations in Chapter 15 work with many of the aspects of both experimental and theoretical probability. Specifically,

- Explorations 15.1 - 15.4 work with experimental probability and its connection to the Law of Large numbers.
- Explorations 15.5 - 15.9 look at theoretical probability, its properties, and different ways to compute it.
- Explorations 15.10 - 15.12 consider other topics associated with probability such as odds, expected value, and counting techniques.

Exploration 15.1 Probability as a Measure of Chance

Purpose: Students answer questions to help them understand probability as a measure of chance.

Materials: A deck of cards, a six-sided die, a compass, and unlined paper

Part A: Using Intuitive Words to Understand Probability

A **probability** is a measure of the likelihood that an uncertain event will occur. If the likelihood of an event is not good, we can use intuitive words such as *unlikely*, *improbable*, or *impossible* to describe its probability. If the likelihood is good, we might instead use words such as *likely*, *probable*, or even *certain*.

1. Consider the experiment of drawing a single card from a standard deck of 52 cards. Describe each event as impossible, unlikely, likely, or certain.
   a. Drawing a black card
   b. Drawing a heart
   c. Drawing an ace
   d. Drawing a green card
   e. Drawing a face card
   f. Drawing a red or black card

2. Consider the experiment of rolling a regular six-sided die. Give a set of numbers from the die that represents an event that is
   a. impossible.   b. unlikely.   c. likely.   d. certain.

3. Use a compass to draw six circles on a sheet of paper. Suppose that each circle represents a spinner. For each question, divide and label a spinner with sections that make the given condition true.
   a. The numbers 1, 2, 3, and 4 are equally likely.
   b. The spinner is most likely to stop on 3.
   c. The spinner is unlikely to stop on 2.
   d. The spinner cannot stop on 1.
   e. The spinner will stop on 4.
   f. The spinner is likely to stop on 2 or 3, but not 1 or 4.

4. In general, how did you determine
   a. which word to use to describe the probability of any one event in Question 1?
   b. how many numbers to include in each set for Question 2?
   c. how large to make the sectors on each spinner in Question 3?

5. Do your answers to Question 4 suggest a way to numerically describe whether an event is more or less likely to occur? Explain.

Part B: Using Numbers to Understand Probability

Because probability is a measure, it can be assigned a numerical value. Specifically, the probability of an event is the ratio of the favorable outcomes to the total number of outcomes or repetitions of an experiment.

1. Give a possible interpretation for each probability statement.
    a. The chance of rain tomorrow is 40%.
    b. There is a 1 in 3 chance that I will need a substitute teacher tomorrow.
    c. The probability of Mya winning the game is $\frac{2}{5}$.

2. Consider the experiment of rolling a six-sided die. Give a set of numbers from the die that has the given probability.
    a. $\frac{1}{6}$ b. $\frac{1}{2}$ c. $\frac{1}{3}$ d. 0   e. 1

3. Consider an event for an experiment with the given probability. Describe the event as impossible, unlikely, likely, or certain.
    a. $\frac{2}{3}$ b. $\frac{1}{20}$ c. $\frac{0}{6}$ d. $\frac{99}{100}$ e. $\frac{10}{10}$

4. If an event is less likely to occur, will the number for the probability be big or small? What if the event is more likely to occur?

5. Are there any limitations to the numerical values for a probability? Explain.

Part C: Summary and Connections

1. How are probabilities similar to other measures? How are they different?

2. Look for different probability statements on the Internet, in newspapers, or in magazines and journals. Give a possible interpretation for each probability statement you find.

3. a. At what grade level do you think students can use intuitive words to describe probabilities? What about describing probabilities with numbers?
   b. What prerequisite knowledge do students need to know before they can begin to describe probabilities numerically?

Exploration 15.2 Testing the Law of Large Numbers

Purpose: Students test the Law of Large Numbers through direct experimentation.

Materials: Coins, six-sided dice, and a computer with Internet access

Part A: Testing the Law of Large Numbers with a Coin

The Law of Large Numbers states that as the number of trials for a given experiment increases, the variation in the experimental probability of a particular event decreases until it closely approximates a fixed number. To see the Law of Large Numbers in practice, consider an experiment of flipping a coin.

1. Flip a coin 50 times and record the number of heads after 10, 20, 30, 40, and 50 flips. Describe the variation you see in the percent of heads. As the number of heads increases, do the percents tend towards a particular value? If so, what value?

Number of flips	Number of heads	Percent heads
10		
20		
30		
40		
50		

2. Flip the coin 50 more times and record the number of heads after 60, 70, 80, 90, and 100 flips. Is there a change in how the percents vary? If so, how?

Number of flips	Number of heads	Percent heads
60		
70		
80		
90		
100		

3. What would happen to the variation in the percents if you flipped the coin 500 times? 1,000 times? Search the Internet for a website that will flip a coin electronically. Test your conjecture by using the software flip a coin a large number of times. Record the results.

4. a. Based on your results for the three previous questions, what is the probability of getting a heads when a fair coin is flipped?
   b. If you think about the possible outcomes when flipping a coin, why does this probability make sense?

Part B: Testing the Law of Large Numbers with a Die

We can also test the Law of Large Numbers with a die.

1. Roll a six-sided die 60 times and record the number of 1s, 2s, 3s, 4s, 5s, and 6s that are rolled. Calculate the percents for the total number of times that each number was rolled. Describe the variation you see in the percents. Do the percents tend towards a particular value? If so, what value?

Outcome	One	Two	Three	Four	Five	Six
Number of times rolled						
Percent of total						

2. Roll a six-sided die 60 more times and record the number of 1s, 2s, 3s, 4s, 5s, and 6s that are rolled. Recalculate the percents for the total number of times that each number was rolled. Is there a change in how the percents vary? If so, how?

Outcome	One	Two	Three	Four	Five	Six
Number of times rolled						
Percent of total						

3. What do you think would happen to the variation in the percents if you rolled the die 500 times? 1,000 times? Search the Internet for a website that will roll a die electronically. Test your conjecture by having the software roll the die a large number of times. Record the results.

4. a. Based on your results for the three previous questions, what is the probability of rolling any one number on a single roll of a die?
   b. If you think about the possible outcomes when rolling a die, why does this number make sense?

Part C: Summary and Connections

1. The percents in this exploration are experimental probabilities. How do the experimental probabilities of flipping a coin or rolling a die compare to theoretical probabilities for the same experiments?

2. How might the idea behind the Law of Large Numbers be useful when selecting a sample in a statistical study?

Exploration 15.3 What's in the Bag?

Purpose: Students use experimental probability to make predictions.

Materials: Cubes of two different colors and a paper bag

Part A: Play "What's in the Bag?"

Get together with one of your peers to play the game "What's in the bag?". Decide who will be Player 1 and who will be Player 2. Player 1 begins the game by choosing ten cubes from a collection of cubes of two colors. The cubes can be any combination of the two colors. For example, Player 1 may choose 2 red and 8 blue, 4 red and 6 blue, or just ten red. The cubes are placed in a paper bag without Player 2 knowing which cubes have been chosen. Without looking, Player 2 draws a cube from the bag and records the color. The cube is then replaced. The game continues with Player 2 drawing five more cubes, each time recording the color and then replacing the cube. After the sixth draw, Player 2 guesses how many cubes of each color are in the bag. Player 2 then makes four more draws, recording the color of each cube. Player 2 is then allowed to stay with the original guess or make a change to the guess.

- If Player 2's guess is correct after 6 draws and he stays with the correct guess, 4 points are earned.
- If Player 2's guess is incorrect after 6 draws but he guesses correctly after the tenth draw, 2 points are earned.
- If Player 2's final guess is incorrect, 0 points are earned.

Players switch roles and a new color combination is chosen. After three rounds, the person with the most points is the winner.

Part B: Summary and Conclusions

1. How was experimental probability used to make predictions when playing the game?

2. Do you think it is reasonable to play this game in the elementary or middle grades? If so, at what grade level?

Exploration 15.4 Simulations

Purpose: Students use simulations to compute experimental probabilities.

Materials: Paper clip, pencil, six colored tiles (4 of one color and 2 of another)

Part A: Shooting Free Throws

Liam is a basketball player who makes 60% of the free throws he attempts. His team is losing a game by one point, and Liam has just been fouled for a one-and-one situation. That is, he has the chance to get a second shot but only if he is successful on the first shot. Work through the following simulation to determine if Liam's team is most likely to win, lose, or go into overtime because of a tie.

1. The following spinner can be used to simulate the situation. Why is this particular spinner an appropriate choice to model the situation?

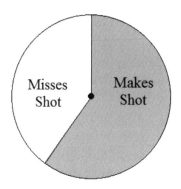

2. Define a trial for the simulation. Be sure to consider what happens if he makes or misses the first shot.

3. Use the tip of your pencil and the paper clip to make the spinner. Conduct 25 trials of the experiment and record your results. Find the experimental probability for each outcome. Based on your results, what is the most likely outcome of the game?

Points Earned	Frequency	Experimental Probability
0		
1		
2		

4. What can you do to be more confident in your choice for the outcome of the game?

Part B: To Replace or Not to Replace? That is the Question!

Henry has a jar of gumballs that contains 4 grape gumballs and 2 orange gumballs. Katie, his little sister, has asked Henry for a gumball. Henry loves to aggravate Katie, so he tells her she must reach into the jar without looking and select a gumball. She can keep the gumball as long as it is not his favorite flavor but she has to select another if her first choice is his favorite. Henry tells her she has two tries to get a gumball. Unbeknownst to Katie, Henry will claim that the first gumball she draws is his favorite, regardless of the flavor. When Katie selects a gumball on the second try, she is allowed to keep it only if the flavor is different from the first one selected.

1. Do you think Katie is more likely to keep the second gumball if she selects grape or orange on the first try? Do you think it matters?

2. Simulate the situation by placing 6 colored tiles (4 of one color to represent the grape gumballs and 2 of another to represent the orange gumballs) in a paper bag. Draw a tile from the bag and record its "flavor." Because Henry told Katie to replace the gumball after the first try, return the tile to the bag and draw again. Conduct 10 trials and record the results.

Trial	1	2	3	4	5	6	7	8	9	10
Flavor of first										
Flavor of second										
Keeps gumball?										

   a. Using the 10 trials, what is the probability that Katie keeps the gumball if she selects
      i. grape on the first try?   ii. orange on the first try?
   b. What can you do to get a better estimate of these probabilities?

3. Does it make any difference in the outcome if Henry does not make Katie replace the gumball after the first try? Use the bag of colored tiles to simulate this situation. Conduct 10 trials and record the results.

Trial	1	2	3	4	5	6	7	8	9	10
Flavor of first										
Flavor of second										
Keeps gumball?										

a. Using the 10 trials, what is the probability that Katie keeps the gumball if she
   i. selects grape on the first try?   ii. selects orange on the first try?
b. What can you do to get a better estimate of these probabilities?

4. How do the probabilities with replacement and without replacement compare?

Part C: Summary and Connections

1. Reconsider the simulation in Part A.
   a. How could you have simulated Liam's free-throw shooting ability by drawing tiles from a bag?
   b. How would you have to change the simulation if Liam were a 70% free-throw shooter?

2. Reconsider the simulation in Part B.
   a. How could you have simulated drawing gumballs by using a six-sided die?
   b. How would you have to change the simulation if there were 6 grape and 4 orange gumballs?

3. Why is it reasonable to use a simulation in both of these situations?

Exploration 15.5 Experimental and Theoretical Probabilities

Purpose: Students compare the values of experimental probabilities to the values of theoretical probabilities.

Materials: An 8-sided die, 10 colored tiles (5 red, 3 blue, and 2 green), and a paper bag

Part A: Probabilities Related to an Eight-sided Die

The Law of Large Numbers states that as the number of trials for an experiment increases, the variation in the experimental probability of a particular event will decrease until it closely approximates a fixed number. Ideally, the fixed number will be the theoretical probability of the same event.

1. The **experimental probability** of an event is the ratio of the favorable outcomes for the event to the total number of times an experiment is conducted. Roll an 8-sided die 160 times and record the number of times each number comes up. Calculate the percents for the total number of times that each number was rolled.

Outcome	One	Two	Three	Four	Five	Six	Seven	Eight
Number of times rolled								
Percent of total								

    a. Why do the percents represent experimental probabilities?
    b. Are the percents close to what you would expect? Why or why not?

2. The **theoretical probability** of an event is the ratio of the favorable outcomes for the event to the number of outcomes in the sample space. A **sample space** is the set of all possible outcomes for an experiment.
    a. What are the outcomes in the sample space of rolling an 8-sided die?
    b. How many times does each number occur in the sample space?
    c. Compute the theoretical probability of rolling each number. Write your answer as a percent.

    $P(1) =$ $P(2) =$ $P(3) =$ $P(4) =$

    $P(5) =$ $P(6) =$ $P(7) =$ $P(8) =$

3. a. How do the experimental probabilities you obtained in Question 1 compare to the theoretical probabilities you obtained in Question 2?
    b. How do you think the probabilities would have compared if you had rolled the die 500 times? 1,000 times?

© 2014 Cengage Learning. All Rights Reserved. May not be scanned, copied or duplicated, or posted to a publicly accessible website, in whole or in part.

Part B: Probabilities of Drawing Colored Tiles

1. Place the ten colored tiles (5 red, 3 blue, and 2 green) in a bag. Draw one tile from the bag, record the color, and then put the tile back in the bag. Repeat the experiment 100 times, each time replacing the tile and shaking the bag. Record your results in the following table and calculate the percents for the total number of times that each color was drawn.

Color	Red	Blue	Green
Number of times drawn			
Percent of total			

    a. Why do the percents represent experimental probabilities?
    b. Are the percents close to what you would expect? Why or why not?

2. a. Describe the sample space for the experiment. How many outcomes are in it?
    b. How many times does each color occur in the sample space?
    c. Compute the theoretical probability of drawing each color. Write your answer as a percent.

    $P(R) =$ $P(B) =$ $P(G) =$

3. a. How do the experimental probabilities you obtained in Question 1 compare to the theoretical probabilities you obtained in Question 2?
    b. How do you think the probabilities would have compared if you had drawn a tile 500 times? 1,000 times?

Part C: Summary and Connections

1. If theoretical probabilities are more exact, why do you think it is important to also know how to compute experimental probabilities?

2. Are children more likely to compute theoretical or experimental probabilities in the elementary grades? What about the middle grades? Examine a set of curriculum materials to find out.

Exploration 15.6 Properties of Probability

Purpose: Students use a deck of cards to explore the properties of probability.

Materials: A standard deck of 52 cards

Part A: Mutually Exclusive Events

Take a few moments to become familiar with a deck of cards. A standard deck contains two colors; red and black, and four suits: spades (♠), clubs (♣), hearts (♥), and diamonds (♦). There are also 13 denominations; 2, 3, 4, 5, 6, 7, 8, 9, 10, jack (J), queen (Q), king (K), and ace (A). The jacks, queens, and kings are called face cards because of the faces printed on them. Use a standard deck of cards to complete each activity.

1. Write the sample space for drawing a single card from the deck. How many outcomes are in the sample space?

2. Two events are **mutually exclusive** if they have no outcomes in common. Suppose one card is drawn from the deck. Use the sample space to determine whether each pair of events is mutually exclusive. For those that are not, explain why.

    a. Event A: Drawing a red card
       Event B: Drawing a black card

    b. Event A: Drawing an ace
       Event B: Drawing a heart

    c. Event A: Drawing a 10
       Event B: Drawing a red card

    d. Event A: Drawing a 3
       Event B: Drawing a king

3. Consider the mutually exclusive events of drawing a queen or drawing a 3 when one card is drawn from the deck.
    a. Use the sample space to find the probability of drawing a queen.
    b. Use the sample space to find the probability of drawing a three.
    c. Add the probabilities from parts (a) and (b).
    d. Use the sample space to find the probability of drawing a queen or a three. How does the probability compare to your answer from part (c)? Why does this make sense?

4. Consider the mutually exclusive events of drawing a red card or drawing a spade when one card is drawn from the deck.
    a. Use the sample space to find the probability of drawing a red card.
    b. Use the sample space to find the probability of drawing a spade.
    c. Add the probabilities from parts (a) and (b).
    d. Use the sample space to find the probability of drawing a red card or a spade. How does the probability compare to your answer from part (c)? Why does this make sense?

5. If two events, *A* and *B*, are mutually exclusive, what generalization can you make about their probabilities? State your generalization symbolically.

Part B: Complementary Events

Two events are **complementary** if they are mutually exclusive and every outcome in the sample space can be placed into one event or the other. Use the sample space for drawing a single card from the deck to answer each question.

1. Determine whether each pair of events is complementary. For those that are not, explain why.
   a. Event *A*: Drawing a red card
      Event *B*: Drawing a black card
   b. Event *A*: Drawing a heart
      Event *B*: Drawing a spade
   c. Event *A*: Drawing a card with a letter
      Event *B*: Drawing a card with a number
   d. Event *A*: Drawing a spade
      Event *B*: Drawing a 4

2. Consider the complementary events of drawing a red card or drawing a black card when one card is drawn from the deck.
   a. Use the sample space to find the probability of drawing a red card.
   b. Use the sample space to find the probability of drawing a black card.
   c. Add the probabilities from parts (a) and (b). What value do you get? Why does this value make sense?

3. Consider the complementary events of drawing a diamond or drawing a card that is not a diamond when one card is drawn from the deck.
   a. Use the sample space to find the probability of drawing a diamond.
   b. Use the sample space to find the probability of drawing a card that is not a diamond.
   c. Add the probabilities from parts (a) and (b). What value do you get? Why does this value make sense?

4. a. If two events, *A* and *B*, are complementary, what generalization can you make about their probabilities? State your generalization symbolically.
   b. If you knew the probability of an event *A*, how could you find the probability of its complement?

5. Why must complementary events be mutually exclusive?

Part C: Non-Mutually Exclusive Events

Two events are **non-mutually exclusive** if they have outcomes in common. Use the sample space for drawing a single card from the deck to answer each question.

1. Determine whether each pair of events is non-mutually exclusive. For those that are not, explain why.
   a. Event A: Drawing a red card
      Event B: Drawing a heart
   b. Event A: Drawing a face card
      Event B: Drawing a 9
   c. Event A: Drawing a spade
      Event B: Drawing a diamond
   d. Event A: Drawing an ace
      Event B: Drawing a black card

2. Consider the non-mutually exclusive events of drawing a queen or drawing a black card when one card is drawn from the deck.
   a. Use the sample space to find the probability of drawing a queen.
   b. Use the sample space to find the probability of drawing a black card.
   c. Add the probabilities from parts (a) and (b).
   d. Use the sample space to find the probability of drawing a queen or a black card. How does the probability compare to your answer from part (c)? What causes the discrepancy?

3. Consider the non-mutually exclusive events of drawing a face card or drawing a heart when one card is drawn from the deck.
   a. Use the sample space to find the probability of drawing a face card.
   b. Use the sample space to find the probability of drawing a heart.
   c. Add the probabilities from parts (a) and (b).
   d. Use the sample space to find the probability of drawing a face card or a heart. How does the probability compare to your answer from part (c)? What causes the discrepancy?

4. In Questions 2 and 3, the probability you found by adding was different from the one found directly from the sample space. What can be done to make the two probabilities the same? Why does this make sense?

5. If two events, A and B, are non-mutually exclusive, what generalization can you make about their probabilities? State your generalization symbolically.

Part D: Summary and Connections

1. How can properties of probability make computing probabilities easier?

2. Can your property for mutually exclusive events be generalized to more than two events? If so, write a symbolic statement that does so.

3. Can your property for non-mutually exclusive events be generalized to three events? If so, write a statement that does so.

Exploration 15.7 Multi-Stage Probability

Purpose: Students calculate multi-stage probabilities using a tree diagram.

Materials: Five tiles or blocks (1 green, 2 red, and 2 blue), a bag, and a die

Part A: Experimental Probability for a Game

Get together with one of your peers to play the following game. Put five tiles or blocks in the bag. They should be identical except for color. The game is played by drawing a tile from the bag (without looking) and rolling a die. A win occurs when a blue tile is drawn and an odd number is rolled. For example, a draw of a blue tile and a roll of 5 is a win; however, a draw of a red tile and a roll of 5 is NOT a win.

1. Do you think you are likely to win the game? Why or why not?

2. Play the game 20 times. Record the color of the tile and the number rolled on the die. Be sure to replace the tile once you have recorded the outcome.
   a. How many of the outcomes are winning outcomes? Losing outcomes?
   b. Use your 20 trials to find the experimental probability of winning the game. Does this match your expectations? Why or why not?
   c. What could you do to be more confident in your experimental probability of winning the game?

Part B: Theoretical Probability for the Game

1. Write the sample space for the game you played in Part A. Use the sample space to find the probability of winning the game. How does your experimental probability compare to the theoretical probability?

2. Another way to find the theoretical probability is to write probabilities on a tree diagram. Complete the following tree diagram by placing the probability of each outcome in each stage on the appropriate branch.

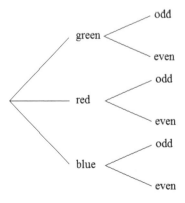

3. Consider the probability on the blue branch followed by the probability on the odd branch.
    a. How are these values related to you answer in Question 1?
    b. Make a conjecture about how to compute a multi-stage probability if the probabilities at each stage are known. Test your conjecture with other outcomes from the game.

Part C: Using a Tree Diagram

Trina's favorite meal is pasta with ice cream for dessert. Trina's mom cooks pasta once a week. If her mom cooks pasta, the probability that Trina has ice cream for dessert is $\frac{2}{3}$. If her mom does not cook pasta, then the probability she has ice cream is $\frac{1}{4}$.

1. Draw a tree diagram that represents the problem. Use it to find the probability of Trina having pasta and ice cream.

2. What other questions can you answer using the tree diagram?

Part D: Summary and Connections

1. Write a story problem that can be solved using a multi-stage probability. How does the language in your problem reflect the fact that a multi-stage probability is to be used?

2. A tree diagram is not always the most efficient way to compute multi-stage probabilities. In what situations would it be difficult to use a tree diagram?

Exploration 15.8 Misconceptions with Probability

Purpose: Students identify common misconceptions associated with learning probability.

Materials: Paper and pencil

Part A: What is the Misconception?

The following statements represent common misconceptions students have when learning about probability. Identify the misconception and then explain how you might correct the student's thinking.

1. Emily got 3 heads when she tossed a coin 3 times. She says she is more likely to get a tail than a head on the fourth flip.

2. A bag contains 4 red blocks and 5 blue blocks. James says that if he picks a block at random, the probability that the block will be red is $\frac{4}{5}$.

3. Benjamin says that if he rolls two dice and finds the sum of the numbers, the probability of getting a sum of 6 is $\frac{1}{11}$.

4. Margaret says that if she rolls six *fair* dice simultaneously, she is more likely to get outcomes of 1, 2, 3, 4, 5, 6 than outcomes of 1, 1, 1, 1, 1, 1.

5. Box *A* has more black balls than Box *B*. Logan says if she chooses 1 ball from each box, she is more likely to choose a black ball from Box *A* than from Box *B*.

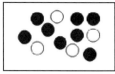
Box *A*

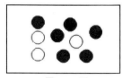
Box *B*

6. Macy says that tomorrow it will either rain or not rain. Therefore, the probability that it will rain is 0.5.

7. Caleb rolled a die twelve times without getting a 4. He says the probability of getting a 4 on his next roll is more than $\frac{1}{6}$.

8. Mia tells her friend that it is foolish to buy a lottery ticket with the numbers 1, 2, 3, 4, 5, and 6 on it because this is less likely to occur than other combinations.

9. Ryan says the probability of getting gray is 50% for each spinner.

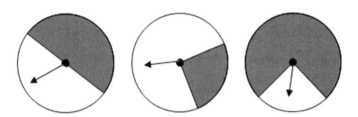

10. A student says his team will win, lose, or tie the football game. Therefore, the probability that his team will win is $\frac{1}{3}$.

11. Joseph is about to toss two pennies. He says that the probability he will get a head and a tail is $\frac{1}{3}$.

12. Mrs. Smith is scheduled for a major operation. Her doctor told her that 85% of the people who have this operation make a complete recovery. Riley says there is an 85% chance that Mrs. Smith will make a complete recovery if she has this operation.

Part B: Summary and Connections

1. Take a look at the mistakes that were made by the students. Are there any similarities or differences in the errors the students made? If so, can you classify the errors?

2. How many of your friends have these common misconceptions with probability? Pose these situations to your friends to find any misconceptions they might have. Write a short report about what you find.

Exploration 15.9 Geometric Probability

Purpose: Students compute geometric probabilities using number lines and areas of geometric figures.

Materials: Paper and pencil

Part A: Using a Number Line to Find a Probability

In many probabilistic situations, the outcomes in the sample space are made of continuous measures rather than discrete objects. In such cases, we can use **geometric probabilities** to find the likelihood of an event by taking the measure of the event to the measure of the sample space. The measures can relate to any measurable attribute such as length or area.

1. A piece of spaghetti 10 in. long is randomly broken into 2 pieces. The following segment shows the piece of spaghetti subdivided into 1-in. pieces. Shade the regions on the segment where a break could occur so that both pieces are at least 3 in. long.

    a. What does the shaded portion of the segment represent in terms of the experiment?
    b. What does the length of the segment represent in terms of the experiment?
    c. What do you get if you take the ratio of the shaded portion to the entire length?

2. Suppose a piece of spaghetti 10 in. long is randomly broken into 2 pieces. Shade the segment where a break could occur so that one piece is less than 3 in. long.

    a. What is the probability that one piece will be less than 3 in. long?
    b. What is the probability that both pieces will be less than 3 in. long?
    c. What is the probability that one piece will be at least 5 in. long?

3. A subway train arrives at a station every 15 minutes. If you arrive at the station at a random time, what is the probability that you will have to wait more than 5 minutes for a train? Use a number line to represent and solve the problem.

Part B: Using Area to Find a Probability

Geometric probabilities can also be computed with areas. Consider the following questions.

1. An unusual dartboard is made from a circle with a diameter of 10 cm placed inside of a square with sides of 10 cm. A dart is thrown at random at the dartboard and lands within the square.

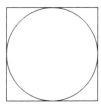

   a. Suppose the dart must hit inside the circle to produce a favorable outcome. What is the area of the circle?
   b. Suppose the square represents the possible outcomes for the experiment. What is the area of the square?
   c. What do you get if you take the ratio of the area of the circle to the area of the square?

2. Suppose a dart is thrown at random and hits the following unusual dartboard.

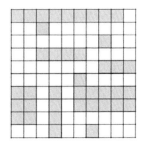

   a. What is the probability that the dart hits a shaded square?
   b. How did you use area to compute the probability?

3. Why is it important that the darts hit the dartboards in the previous two questions?

Part C: Summary and Connections

1. How is the computation of a geometric probability similar to the computation of other theoretical probabilities? How is it different?

2. Write a story problem that involves geometric probability and an angle measure or volume. Give your problem to a partner to solve.

Exploration 15.10 Understanding Odds

Purpose: Students answer questions to understand odds as a measure of chance.

Materials: Paper and pencil

Part A: Interpreting Odds

Odds are another way to express the likelihood that an uncertain event will or will not occur. **Odds in favor** gives the likelihood of an event as the ratio of the number of successful outcomes to the number of unsuccessful ones. **Odds against** gives the ratio of the number of unsuccessful outcomes to the number of successful ones.

1. Give a possible interpretation for each statement involving odds.
    a. The odds in favor of Fast Freddie winning the race are 1 to 4.
    b. The odds in favor of getting a winning number are 1 to 49.
    c. The odds against a rainy day are 3 to 1.
    d. The odds against drawing a blue chip are 5 to 2.

2. Each of the following represents the odds in favor of an event A. Do you think event A is impossible, unlikely, likely, or certain?
    a. $1:10$ b. $4:1$ c. $5:6$ d. $0:1$ e. $1:0$

3. A bag contains red chips and blue chips. There are 4 red chips in the bag and one chip is drawn at random. How many blue chips must be in the bag so that the odds in favor of drawing blue are
    a. $1:1$?   b. $1:4$?   c. $2:1$?   d. $3:2$?

4. A bag contains red chips and blue chips. There are 3 red chips in the bag and one chip is drawn at random. How many blue chips must be in the bag so that the odds
    a. in favor of drawing a blue are $1:1$?
    b. against drawing a blue are $3:2$?
    c. in favor or drawing a red are $1:4$?
    d. against drawing a red are $5:3$?

5. Consider a regular six-sided die. Give a specific set of numbers for an event A so that event A has the given odds.
    a. The odds in favor of A are 1:5   b. The odds in favor of A are 1:2
    c. The odds against A are 1:2   d. The odds against A are 1:1

6. Consider the odds in favor of an event A.
    a. If the first number is zero, what must be true about event A?
    b. If the second number is zero, what must be true about event A?

Part B: Odds and Probability

Odds and probability are closely connected. Use the fact that the odds in favor of an event A is the ratio of the number of successful outcomes to the number of unsuccessful outcomes, and the probability of event A is the ratio of the number of successful outcomes to the total number of outcomes, to answer the following questions.

1. Devise a strategy for converting the odds in favor of an event A to a probability for A. Explain your thinking.

2. Devise a strategy for converting a probability for an event A to the odds in favor of A. Explain your thinking.

3. Use your strategy to convert each of the following odds in favor of event A to the probability that A will occur.
   a. 1 : 2   b. 3 : 5   c. 1 : 10   d. 5 : 2

4. Use your strategy to convert each probability for event A to the odds in favor of event A.
   a. $\frac{1}{3}$ b. $\frac{4}{9}$ c. 80%   d. 0.46

5. If the probability for an event A is smaller than 50%, will the first number in the odds in favor of A be bigger or smaller than the second? Explain your thinking.

6. If the probability for an event A is larger than 50%, will the first number in the odds against A be bigger or smaller than the second? Explain your thinking.

Part C: Summary and Connections

1. In what situations have you encountered odds in your daily life?

2. a. Which is more difficult for you to understand and work with: odds or probability? Can you explain why?
   b. What impact might your answer have on the way you teach odds and probability in the future?

Exploration 15.11 Fair Games

Purpose: Students use experimental probability and expected value to determine whether two games are fair.

Materials: Two six-sided dice

Part A: Even and Odd Products

Get together with one of your peers to play the following game. Roll two dice and multiply the numbers. If the product is even, Player 1 wins. If the product is odd, Player 2 wins. Play several rounds, and then use your experience with the game to answer the following questions.

1. Do your initial impressions of the game lead you to think that it is fair for both players? Why or why not?

2. One way to get a better impression of the fairness of the game is to look at experimental probabilities. Play the game 100 times and record the number of times each player wins in the following table. Then compute the experimental probabilities.

Outcome	Frequency	Experimental Probability
Even product (Player 1 wins)		
Odd product (Player 2 wins)		

    a. Based on the experimental probabilities, do you think the game is fair for both players? Why or why not?
    b. What can you do to be more confident in your decision about the fairness of the game?

3. Expected value can also be used to determine whether a game is fair.
    a. When two dice are rolled, what is the theoretical probability that the product of the two numbers is even? That the product is odd?
    b. The only outcomes in the game are a win or a loss. What are reasonable point values to assign to these outcomes?
    c. Use your answers to previous questions to find the expected value of the game. Is the game fair?

Part B: Rock, Paper, Scissors

Rock, paper, scissors is a common children's game that uses simple hand motions. Two players start by making a fist and then at the count of three, make one of three gestures; rock by showing a fist, paper by showing an open hand, and scissors by showing two fingers spread apart. If the players make different gestures, then rock beats scissors, scissors beat paper, and paper beats rock. If the players make the same gesture, it is a tie. Play several rounds and use your experience with the game to answer the following questions.

1. Do your initial impressions of the game lead you to think that it is fair for both players? Why or why not?

2. Play the game 100 times and record the number of times each outcome occurs. Then compute the experimental probabilities.

Outcome	Frequency	Experimental Probability
Player 1 wins		
Player 2 wins		
Tie		

    a. Based on the experimental probabilities, do you think the game is fair for both players? Why or why not?
    b. Unlike rolling dice, the players have some control of the outcomes. How can this affect the fairness of the game?

3. If it is assumed that players select a gesture completely at random, then expected value can be used to determine the fairness of the game.
    a. Use a tree diagram to find the possible outcomes and their respective probabilities.
    b. What is a reasonable point value to assign to a win? A loss? A tie?
    c. Use your answers to previous questions to find the expected value of the game. Is the game fair?

Part C: Summary and Connections

1. Why is it important to know about fairness when playing games of chance?

2. Expected value can be used in a variety of situations other than with games of chance. Do an Internet search for different applications of expected value. Make a list of the applications you find.

Exploration 15.12 Pascal's Triangle and Combinations

Purpose: Students make connections between Pascal's triangle and combinations.

Materials: A calculator

Part A: Pascal's Triangle

Pascal's triangle is an infinite triangular array of numbers that can be generated in the following way. Begin with three ones arranged in a triangular fashion. Add rows to the bottom of the triangle using the following rules:

- Every row begins and ends with a one.
- Entries between the ones are generated by adding the two entries that are diagonally above in the preceding row.

For instance, the first four rows of Pascal's triangle are as follows:

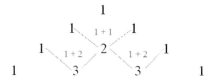

Generate the first ten rows of Pascal's triangle. What patterns do you see?

Part B: Comparing Combinations to Pascal's Triangle

1. Compute the values of $_2C_2$, $_2C_1$, and $_2C_0$. Compare them to the third row of Pascal's triangle. What do you notice?

2. Compute the values of $_4C_4$, $_4C_3$, $_4C_2$, $_4C_1$, and $_4C_0$. Compare them to the fifth row of Pascal's triangle. What do you notice?

3. Make a conjecture about how you can use Pascal's triangle to compute combinations. Test several values to see if your conjecture holds.

4. Use Pascal's triangle to compute the following values.
   a. $_5C_3$ b. $_6C_2$ c. $_7C_5$ d. $_8C_4$ e. $_9C_5$

Part C: Summary and Connections

1. What advantages are there to using Pascal's triangle to compute combinations?

2. Can you use Pascal's triangle to compute probabilities? If so, how?

Appendix A
Black Line Masters

Appendix A contains the black line masters for the explorations in this manual. These masters may be copied for classroom use only. They include:

- A decimal place value mat
- 1-in. grid paper
- Quarter-inch grid paper
- Rectangular dot paper
- 5 × 5 geoboard dot paper
- 11 × 11 geoboard dot paper
- An Equation mat
- Half-inch grid paper
- Decimal grid paper
- Isometric dot paper
- 6 × 6 geoboard dot paper
- A table of random digits

Decimal Place Value Mat

Thousands	Hundreds	Tens	Ones

Equation Mat

One-Inch Grid Paper

Half-Inch Grid Paper

Quarter-Inch Grid Paper

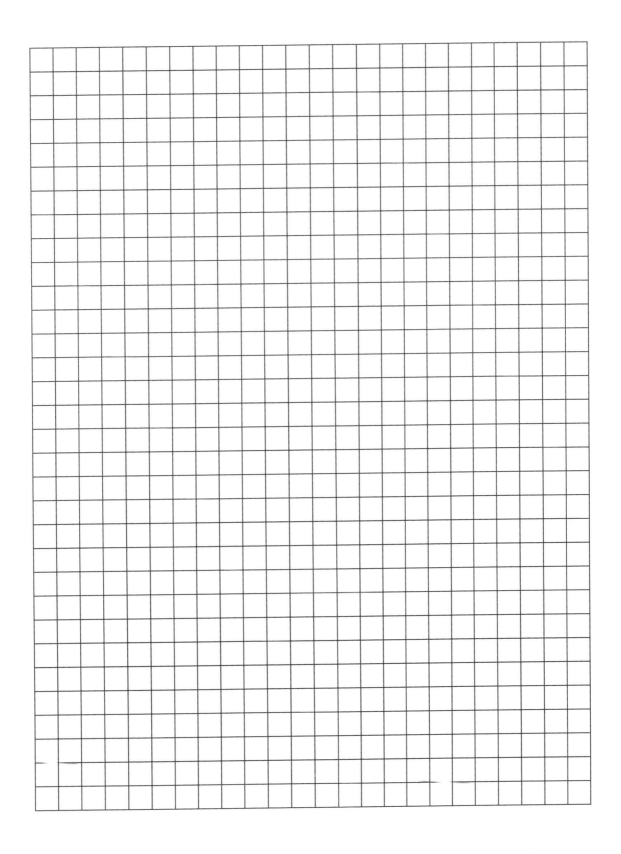

Decimal Grid Paper

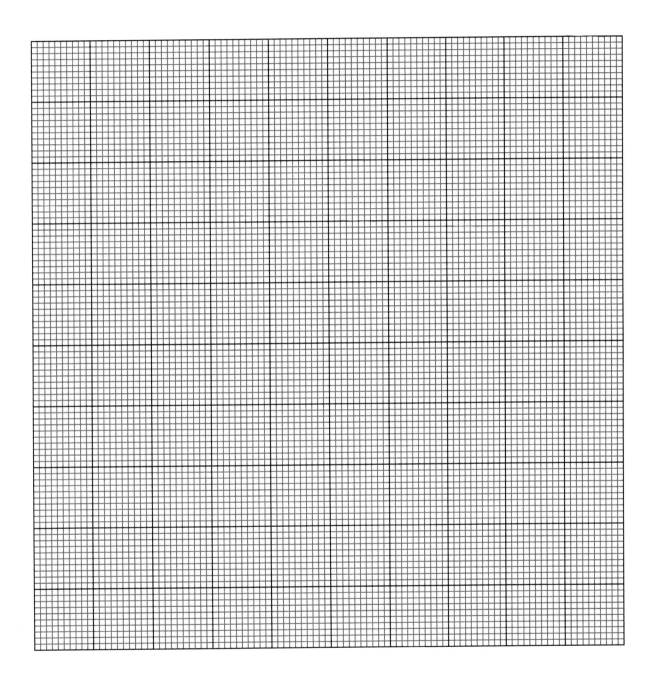

Rectangular Dot Paper

Isometric Dot Paper

5 × 5 Geoboard Paper

A-10

6 × 6 Geoboard Paper

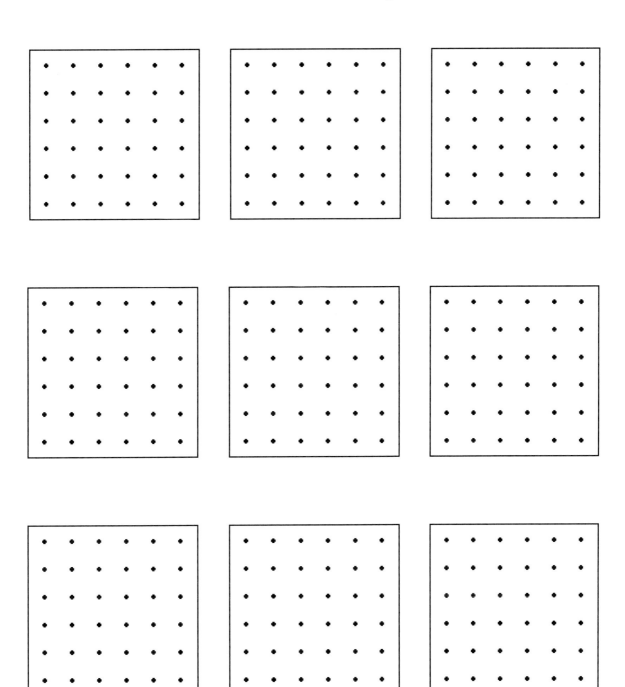

11 × 11 Geoboard Paper

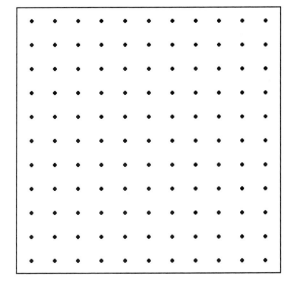
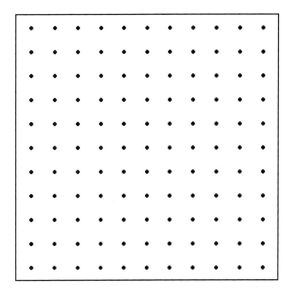
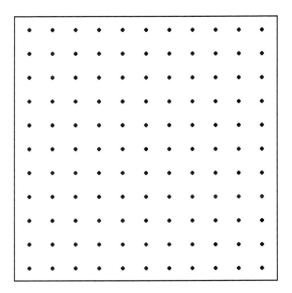
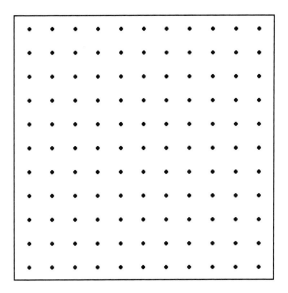

Table of Random Digits

26907 88173 71189 28377 13785 87469 35647 19695 33401 51998
86668 70341 66460 75648 78678 27770 30245 44775 56120 44235
04982 68470 27875 15480 13206 44784 83601 03172 07817 01520
28549 98327 99943 25377 17628 65468 07875 16728 22602 33892
11762 54806 02651 52912 32770 64507 59090 01275 47624 16124

02805 52676 22519 47848 68210 23954 63085 87729 14176 45410
36116 42128 65401 94199 51058 10759 47244 99830 64255 40516
55216 63886 06804 11861 30968 74515 40112 40432 18682 02845
36248 36666 14894 59273 04518 11307 67655 08566 51759 41795
12386 29656 30474 25964 10006 86382 46680 93060 52337 56034

22784 07783 35903 00091 73954 48706 83423 96286 90373 23372
07330 07184 86788 64577 47692 45031 36325 47029 27914 24905
22565 02475 00258 79018 70090 37914 27755 00872 71553 56684
06644 94784 66995 61812 54215 01336 75887 57685 66114 76984
44882 33592 66234 13821 86342 00135 87938 57995 34157 99858

19082 13873 07184 21566 95320 28968 31911 06288 77271 76171
61869 33093 81129 06481 89281 83629 81960 63704 56329 10357
49333 78482 36199 11355 86044 88760 03724 22927 91716 92332
38746 81271 96260 98137 60275 22647 33103 50090 29395 10016
93369 13044 69686 78162 29132 51544 17925 56738 32683 83153

Appendix B
Manipulatives

Appendix B contains many of the manipulatives used throughout this manual. Specifically, it includes:

- Algebra tiles
- Base-ten blocks
- Counters
- Fraction disks
- Pattern blocks
- Tangrams

- Attribute blocks
- Colored rods
- Fraction bars
- One-Inch Square Tiles
- Pentominoes
- Two-colored chips

Algebra Tiles

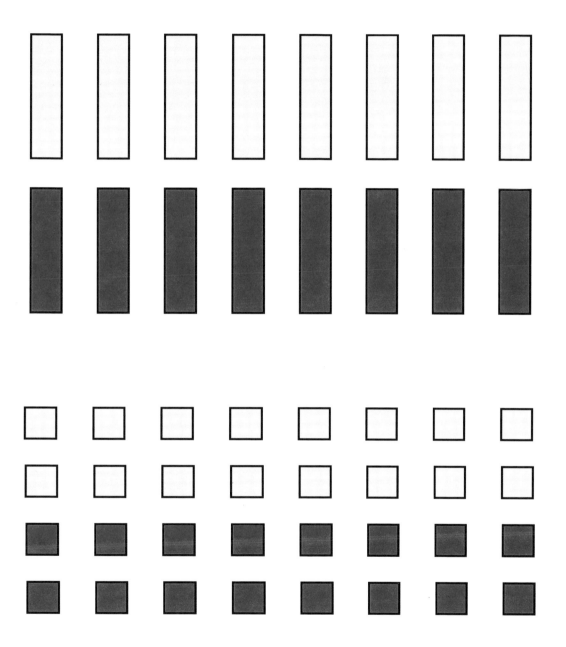

Attribute Blocks

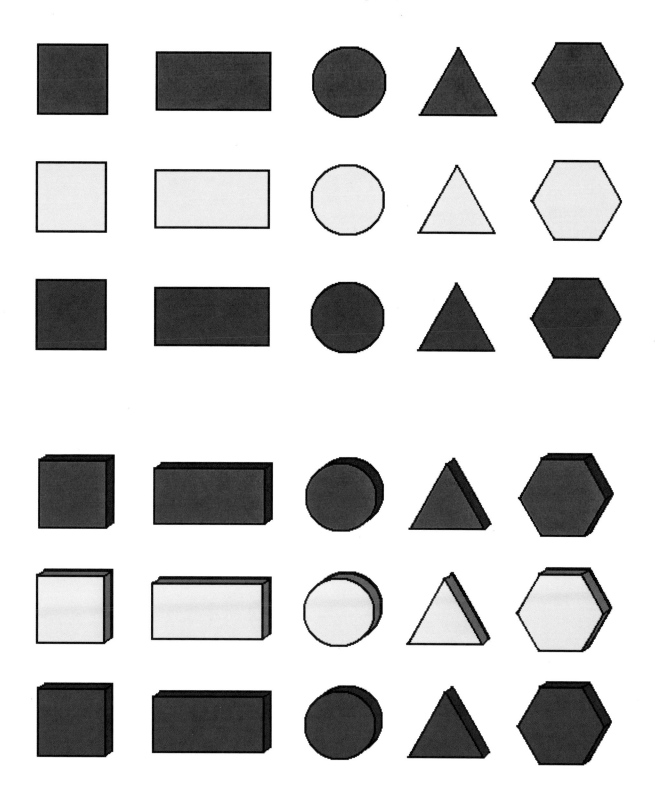

B-3

Attribute Blocks

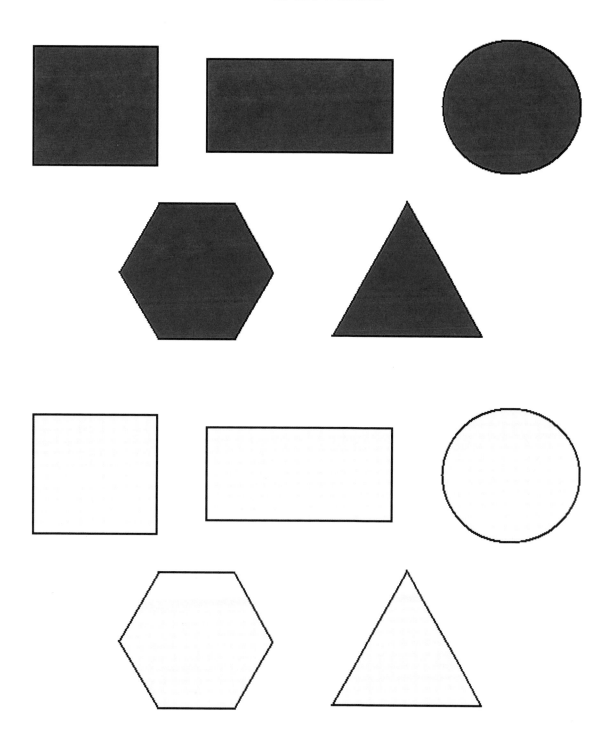

Attribute Blocks

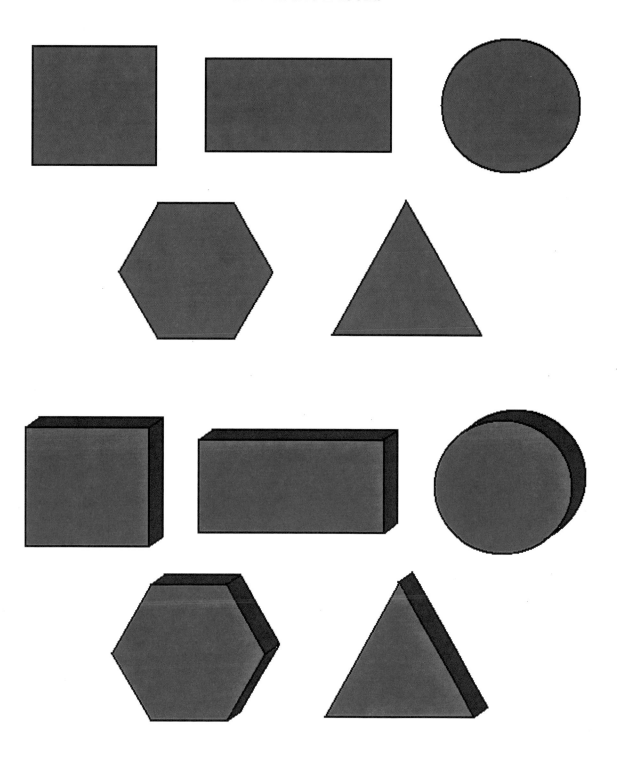

Attribute Blocks

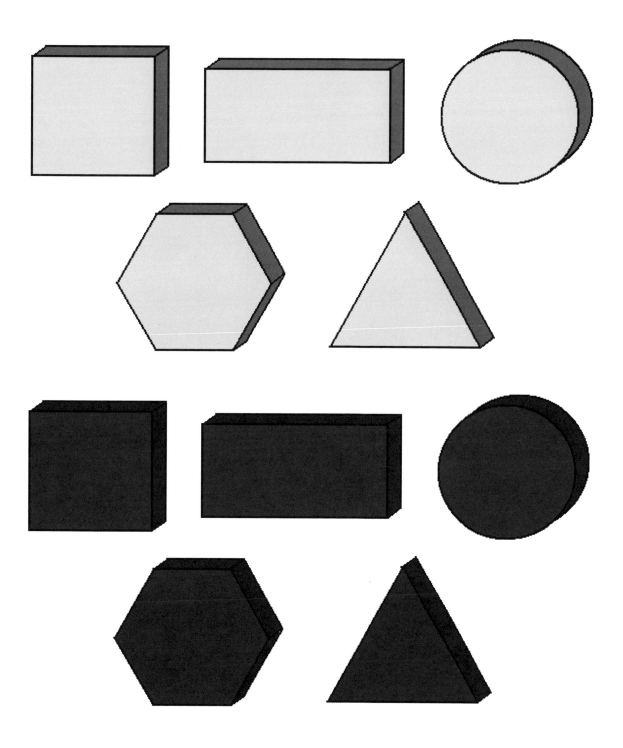

Base-Ten Blocks

Base-Ten Blocks

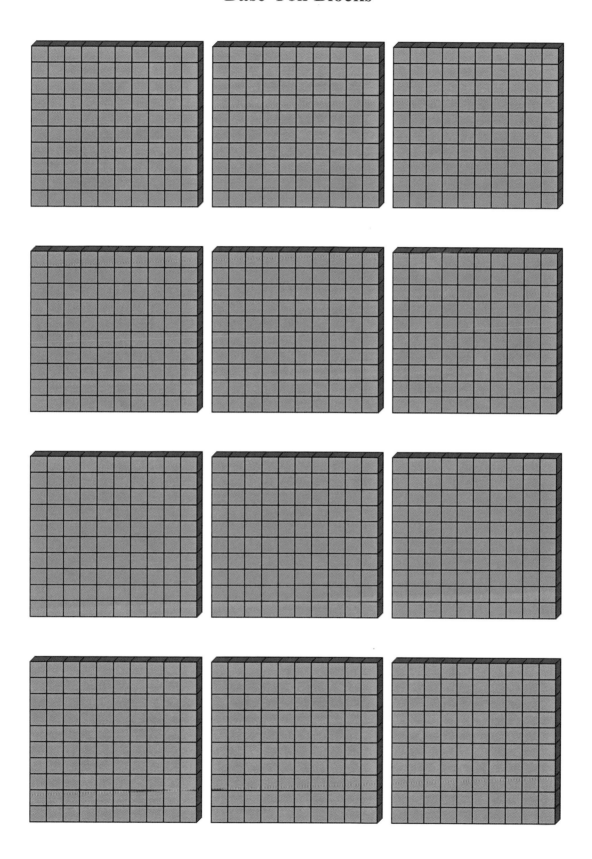

Base-Ten Blocks

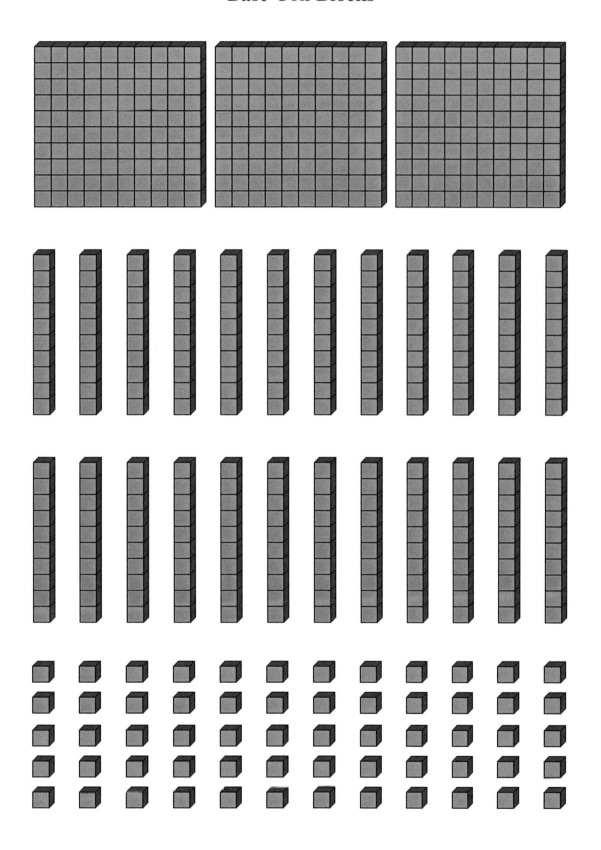

Colored Rods

Counters

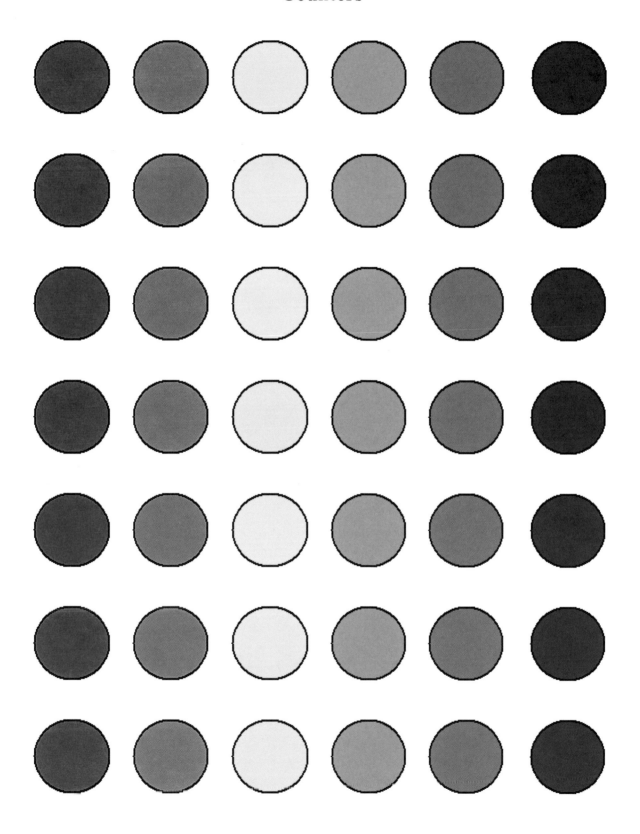

Fraction Bars

1

| 1/2 | 1/2 |

| 1/3 | 1/3 | 1/3 |

| 1/4 | 1/4 | 1/4 | 1/4 |

| 1/5 | 1/5 | 1/5 | 1/5 | 1/5 |

| 1/6 | 1/6 | 1/6 | 1/6 | 1/6 | 1/6 |

| 1/8 | 1/8 | 1/8 | 1/8 | 1/8 | 1/8 | 1/8 | 1/8 |

| 1/10 | 1/10 | 1/10 | 1/10 | 1/10 | 1/10 | 1/10 | 1/10 | 1/10 | 1/10 |

| 1/12 | 1/12 | 1/12 | 1/12 | 1/12 | 1/12 | 1/12 | 1/12 | 1/12 | 1/12 | 1/12 | 1/12 |

Fraction Disks

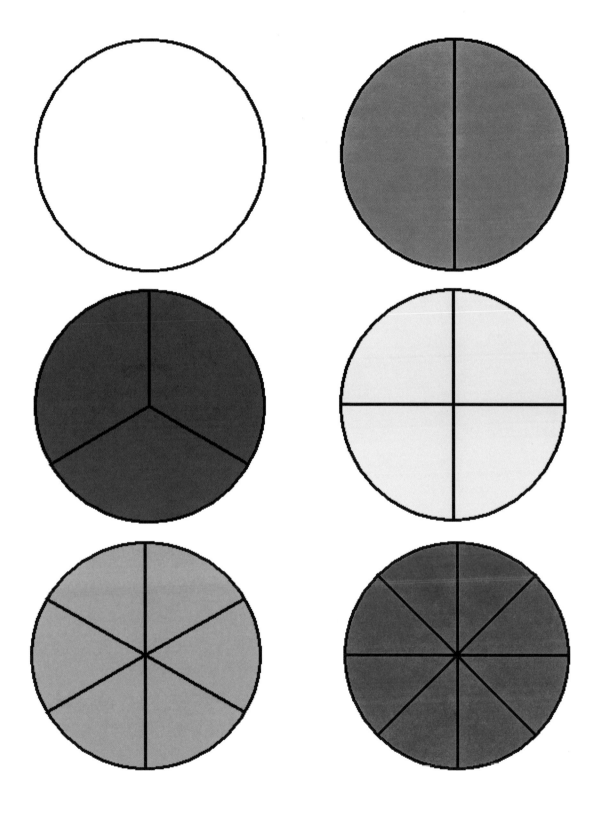

One-Inch Square Tiles

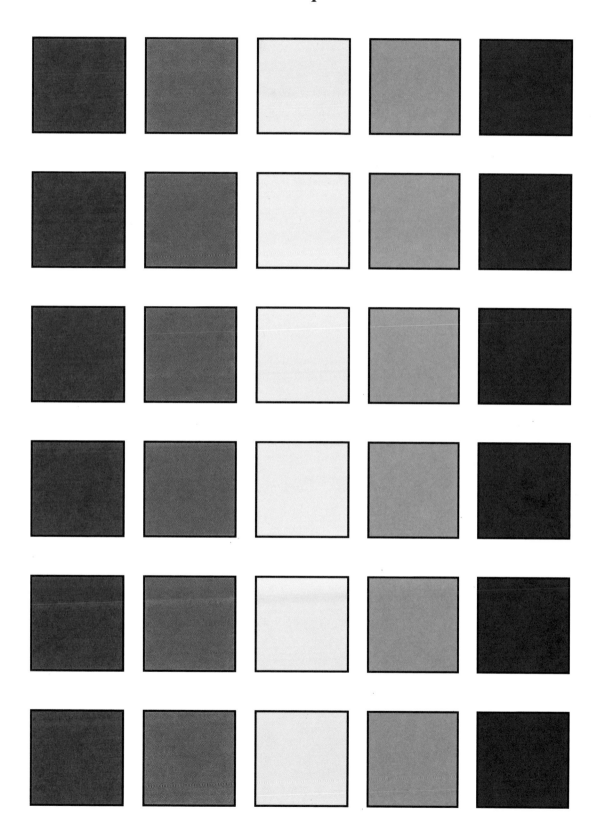

Pattern Blocks

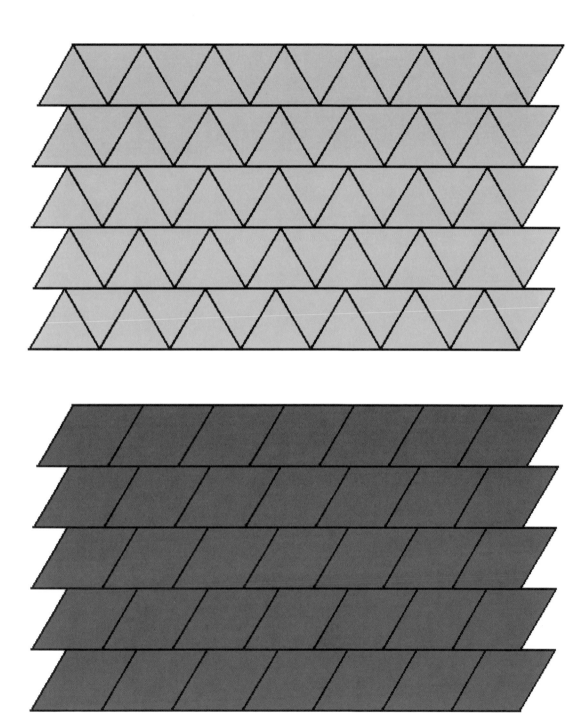

Pattern Blocks

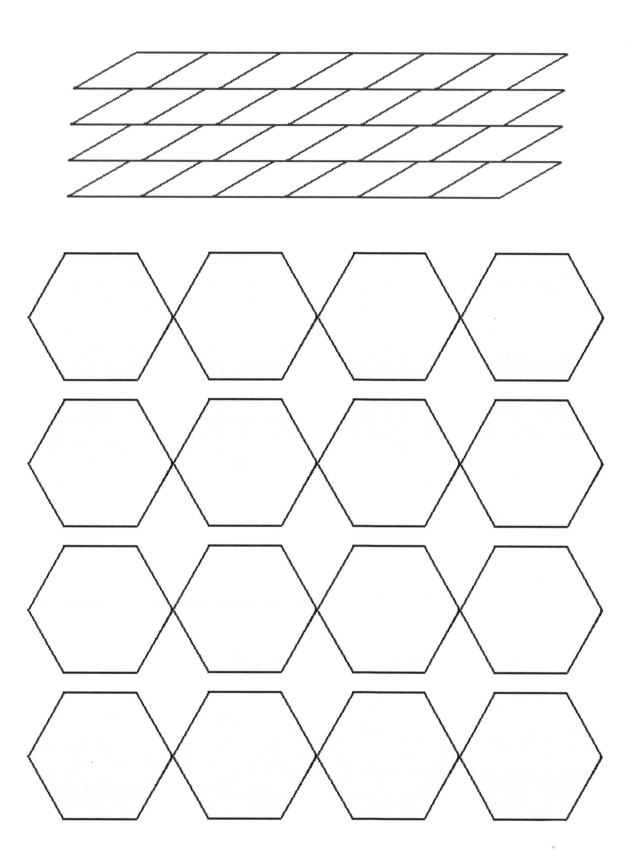

B-16

Pattern Blocks

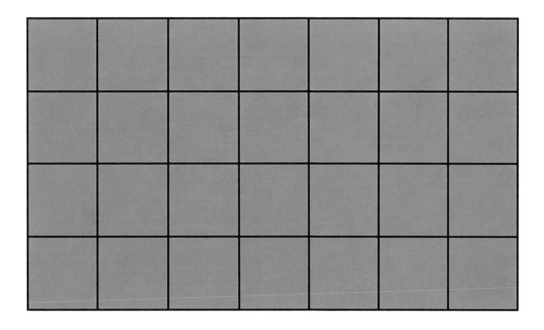

Pentominoes

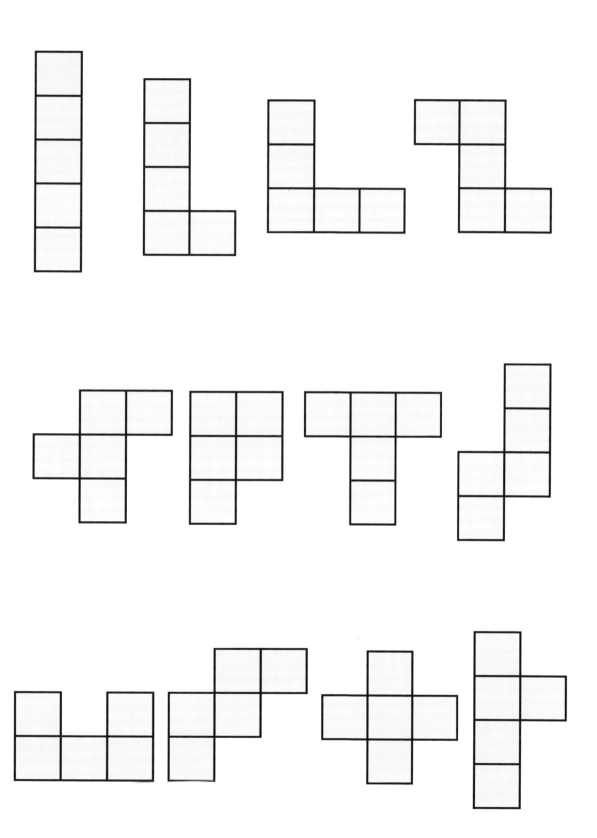

B-18

Tangrams

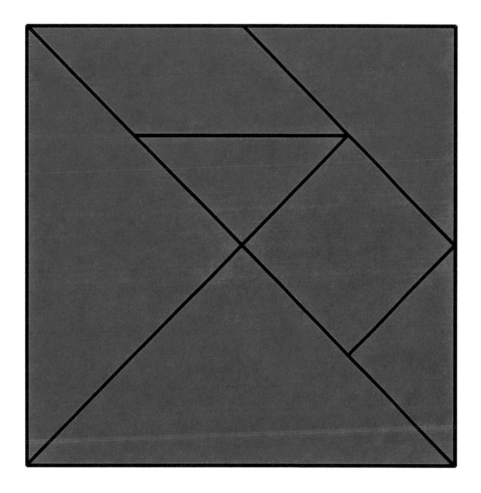

Two-Colored Chips

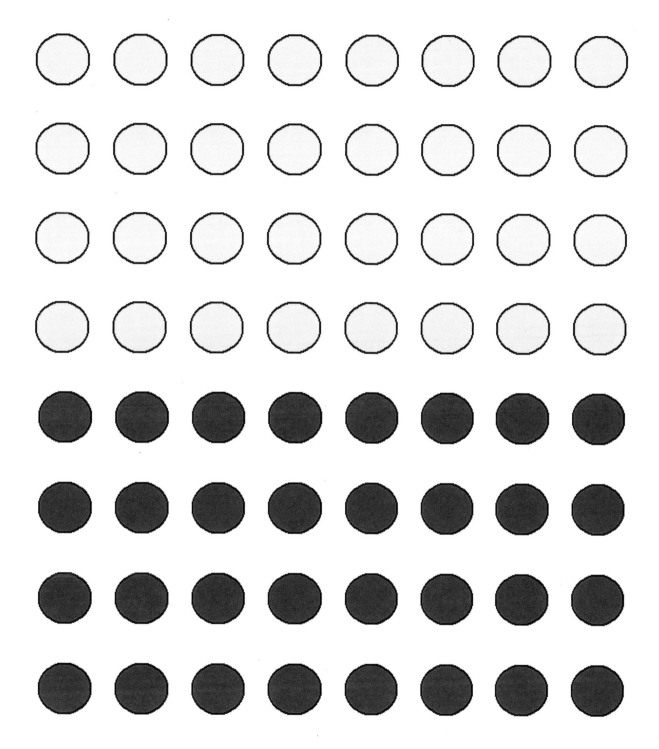